Good Faith Collaboration

History and Foundations of Information Science

Edited by Michael Buckland and Jonathan Furner

Human Information Retrieval by Julian Warner

Good Faith Collaboration: The Culture of Wikipedia by Joseph Michael Reagle Jr.

Good Faith Collaboration: The Culture of Wikipedia

Joseph Michael Reagle Jr.
Foreword by Lawrence Lessig

The MIT Press
Cambridge, Massachusetts
London, England

For information about special quantity discounts, please email special_sales@mit-press.mit.edu

This book was set in Stone Sans and Stone Serif by the MIT Press. Printed and bound in the United States of America.

Library of Congress Cataloging-in-Publication Data

Reagle, Joseph Michael.
Good faith collaboration : the culture of Wikipedia / Joseph Michael Reagle Jr. ; foreword by Lawrence Lessig.
p. cm. — (History and foundations of information science)
Includes bibliographical references and index.
ISBN 978-0-262-01447-2 (hardcover : alk. paper) 1. Wikipedia. 2. Electronic encyclopedias—Case studies. 3. Wikis (Computer science)—Case studies. 4. Communication in learning and scholarship—Technological innovations—Case studies. 5. Authorship—Collaboration—Case studies. 6. Online social networks—Case studies. I. Title.
AE100.R43 2010
030—dc22

2009052779

10 9 8 7 6 5 4 3 2 1

To my family, a barnstar.

Contents

Foreword

There is value in studying anything that was once thought impossible but then proves possible. There is significant value in studying it well. A decade ago, no one—including its founder, Jimmy Wales—would have imagined "Wikipedia" possible. Today it is one of the very top Web sites on the Internet. And not just the Internet: Wikipedia has come to define the very best in an ethic of a different kind of economy or community: at its core, it is a "collaborative community" that freely and voluntarily gives to the world a constant invitation to understand and correct. More than any democracy, it empowers broadly. More than any entity anywhere, it elicits the very best of an amateur ethic—people working hard for the love of the work, and not for the money.

Most of the world has known of Wikipedia for no more than a few years. Even the "digerati" have not paid much attention to the project for more than seven years. Like the most important innovations throughout human history, this one too stole upon us when most of us were looking elsewhere. And now, none of us understands anything new without first pinging Wikipedia's brain to see its cut on whatever piques our curiosity.

Scholars will spend a generation understanding its birth and growth. There have already been important books understanding open source production specifically (Steven Weber, *The Success of Open Source* [Harvard 2004]), and the culture of commons-based production (Yochai Benkler, *The Wealth of Networks* [Yale 2007]).

But Joseph Reagle's contribution here is something new and important. Reagle came to this subject as a native. He was a computer scientist at MIT. He helped me work through early thoughts about what he called "social protocols"—an explicit mixing of computer science ideals with insight about social organization and norms. When he decided to return to graduate school to get his PhD, I was skeptical that such enormous talent should be lost to the stacks for so many years.

This book proves me wrong. Reagle comes to this ethnographical project understanding more about the technology and its history than the people he intended to study. But that knowledge doesn't get in the way. He has opened himself to a community that is similar to some he has worked within—the World Wide Web Consortium, most prominently—but importantly distinct. And as his book convincingly demonstrates, it is a community with a family resemblance to lots in our world, but unlike almost anything else.

Wikipedia is a community, but one formed through a practice, or a doing—collaboration. That collaboration happens within a culture, or a set of norms, guided by principles that the community accepts and fights about, and through that struggle defines. The collaboration produces a social good that an enormous number of people from around the world rely upon. The project is a generation away from its objective of "a world in which every single human being can freely share in the sum of all knowledge." But it is the first time in ten generations that this aspiration of the Enlightenment seems even possible to anyone but the likes of Jefferson.

We need many academic disciplines—economics, political science, history, even law—to help us understand this phenomenon. But the first rich understanding must come from ethnographies. Only a deep reading of the culture of this community—for it is a community rich with a distinctive culture—can begin to make the important lessons of Wikipedia accessible.

No utopia is to be found in these pages. Wikipedia is not written by angels; nor does its founder pretend to perfection. What is most striking throughout this lucid and informed account is the human-ness of everything inside. Wales, the founder, self-consciously practicing the humility every great leader teaches. A community, struggling to get it right, some devoting thousands of hours to making knowledge free.

There are relatively few organizations that inspire respect, flaws notwithstanding. Very few retain that respect after serious scrutiny. These pages introduce one such institution. No one doubts it produces an encyclopedia that has errors. But it is hard to imagine a more significant and sustained community, manned by volunteers, from teenagers to retirees, working to produce understanding.

Every serious soul must try to understand this impossibility. For there is little doubt that its lessons have much to teach far beyond the millions of entries on Wikipedia pages. Nor that an important first step in that understanding is found in these pages.

Lawrence Lessig
Professor of Law, Harvard Law School
Director, Edmond J. Safra Foundation Center for Ethics

Preface

When I returned to graduate school I faced the task of choosing a research topic. I left my position at the World Wide Web Consortium (W3C) with an interest in new Web applications, particularly blogs, and an appreciation for collaboration. Since I had been out of school for a while, and matriculated in an alien discipline, I made frequent use of a new online reference work: Wikipedia. As I grew disenchanted with the fractious narcissism sometimes encountered on blogs, I became increasingly intrigued with the efforts of the individuals producing Wikipedia. While one can find plenty of arguments in both the blogosphere and at Wikipedia, the intention and spirit of the discussion in the two communities is often different. Furthermore, my interest in Wikipedia dovetailed with an especially useful bit of advice I received: the work for a book such as this is long and solitary, so choose something one can live with as it will be the foremost topic on one's mind for years to come. Attempting to understand and portray the spirit of Wikipedia collaboration turned out to be a rewarding obsession.

When I worked at the W3C I participated in and facilitated many working groups. Most people wouldn't think long technical discussions in committees could be very exciting—they usually aren't—but I was inspired by instances in which the varied skills and personalities of my peers complemented one another. The differences that sometimes irritated me or frustrated progress also, sometimes, yielded an elegant solution that exceeded the sum of our individual efforts. I witnessed something similar at Wikipedia. Most people wouldn't think the production of an encyclopedia could be very exciting either, but I found Wikipedia to be a compelling site for the study of collaboration and an endearing but unruly character in a longer historical tale.

Consequently, in the following pages I present a historically informed ethnography of Wikipedia. I also observe that building an encyclopedia is a cumulative and interdependent activity and writing this book is no different. In fact, it seems impossible to properly acknowledge all those who have influenced and supported this work. But as in any other seemingly impossible task, like creating a global encyclopedia, one must start somewhere—and the perfect is the enemy of the good.

I am indebted to Helen Nissenbaum, Gabriella Coleman, and Natalia Levina for their insight, time, and guidance. Thank you. I'm also grateful to other faculty members who helped me while I was at New York University including JoEllen Fisherkeller, Alex Galloway, Brett Gary, Ted Magder, Siva Vaidhyanathan, and Jonathan Zimmerman. When I worked in Cambridge, Massachusetts, I enjoyed perusing the bookshelves of the MIT Press on my lunch breaks, so I'm pleased that this book will appear on those shelves with the help of Marguerite Avery, Susan Clark, Julia Collins, Mel Goldsipe, Emily Gutheinz, Erin Mooney, Johna Picco, and Sharon Deacon Warne. Also, of course, I thank Lawrence Lessig for his support and writing the foreword to this book.

Scholars, colleagues, and friends who discussed drafts, sent comments, listened to me talk an idea through, or pointed out a missed connection or useful reference include Melissa Aronczyk, Phoebe Ayers, Samir Chopra, Shay David, Jonathan Grudin, Said Hamideh, Michael Hart, Sam Howard-Spink, Ian Jacobs, Rob Jones, Jelena Karanovic, Cormac Lawler, Susan Lesch, Lawrence Liang, Andrew Lih, David Parisi, Devon Powers, Evan Prodromou, W. Boyd Rayward, Sage Ross, Aaron Swartz, Michael Zimmer, and Jakob Voss. Michael Buckland, John Broughton, and anonymous reviewers at the MIT Press read the entire manuscript and gave me detailed and patient feedback; this is no small gift. Nora Schaddelee deserves special thanks for putting up with more of my wiki rambling than anyone else, and she also read the earliest—and cruftiest—draft of almost every chapter that follows. Nora, and my brothers Greg and Dan Reagle, were also kind enough to read proofs.

Also, thank you to those who spoke to me about their experiences, especially Wikipedians. I have only named those people who have influenced or commented upon this work specifically, but I benefited from many other conversations. Some appear as sources in this work, but most do not. The vast majority of my sources are a sample of an enormous public discourse, and I portray only a fragment of that. From a methodological perspective it seems odd to thank those who have contributed to a (public) archive and project, but I feel a sense of gratitude nevertheless. Additionally, I'm grateful to my friends and family, including those who might have had little interest in the particulars of this work but wished me well and reminded me that life is larger than the confines of a computer screen.

Finally, I want to recognize a few institutions that in one way or another served as a home to me. The W3C provided much of the inspiration for my interest in collaboration and consensus. The Berkman Center for Internet & Society at Harvard Law School provided me an important opportunity to begin reflecting on the social aspects of online interaction. The Department of Media, Culture, and Communication at the Steinhardt School of New York University generously supported the studies and research that inform this work. Finally, most of this book was written at the Carroll Gardens branch of the Brooklyn Public Library: a hospitable, though often noisy, environment in which I was surprisingly productive up in its little balcony amid comics and tax forms.

1 Nazis and Norms

Show me an admin who has never been called a nazi and I'll show you an admin who is not doing their job.
—J. S.'s Second Law

Wikipedia is not merely an online multilingual encyclopedia; although the Web site is useful, popular, and permits nearly anyone to contribute, the site is only the most visible artifact of an active community. Unlike previous reference works that stand on library shelves distanced from the institutions, people, and discussions from which they arose, Wikipedia is both a community and an encyclopedia. And the encyclopedia, at any moment in time, is simply a snapshot of the community's continuing conversation. This conversation is frequently exasperating, often humorous, and occasionally profound. Most importantly, it sometimes reveals what I call a *good faith* collaborative culture. Wikipedia is a realization—even if flawed—of the historic pursuit of a universal encyclopedia: a technology-inspired vision seeking to wed increased access to information with greater human accord. Elements of this good faith culture can be seen in the following conversation about a possible "neo-Nazi attack" upon the English-language Wikipedia.

In early 2005 members of Stormfront, a "white pride" online forum, focused their sights on Wikipedia. In February they sought to marshal their members to vote against the deletion of the article "Jewish Ethnocentrism," one favored by some "white nationalists" and that made use of controversial theories of a Jewish people in competition with and subjugating other ethnic groups. Stormfront's alert was surprisingly sensitive to the culture of Wikipedia by warning recipients, "you must give your reason as to why you voted to keep the article—needless to say you should do so in a cordial manner, those wishing to delete the article will latch onto anything they can as an excuse to be hostile towards anybody criticising Jewish culture."[1] Six months later, participants of Stormfront, perhaps dissatisfied with their earlier efforts, were considering using the software that runs Wikipedia, or even some of its content, to create their own ("forked") version more to their liking.[2]

The charge of Nazism has a long and odd history in the online community realm. One of the most famous aphorisms from earlier Internet discussion groups is Godwin's Law of Nazi Analogies: "As an online discussion grows longer, the probability of a comparison involving Nazis or Hitler approaches one."[3] Godwin's Law speaks to a tendency of online participants to think the worst of one another. So much so, that the epigraph at the beginning of this chapter, J. S.'s Second Law, implies that if you haven't been called a Nazi, you simply haven't been active enough on Wikipedia.[4] Yet, throughout the immense Wikipedia discussion threads prompted by a potential "neo-Nazi attack" no one compared anyone else to Hitler. Granted, some Stormfront members are self-identified Nazis for whom that term would not be an insult, but there was also serious disagreement among Wikipedians—and even the white racialists reminded themselves they need be cordial on Wikipedia.

This cordiality would be commented upon in a related incident later in 2005, in August, when Wikipedia user Amelkite, the owner/operator of the white supremacist Vanguard-News-Network, had his Wikipedia account blocked. MattCrypto, a Wikipedia administrator, then unblocked him, thinking it unfair to block someone because of his or her affiliation rather than Wikipedia actions. This prompted another administrator, SlimVirgin, to reblock, pointing out Amelkite had posted a list of prominent Wikipedians thought to be Jews as well as information on how to counter Wikipedia controls of disruption. The conversation between Wikipedia administrators remained civil:

MattCrypto: Hi SlimVirgin, I don't like getting into conflict, particularly with things like block wars and protect wars, so I'm unhappy about this. . . .

SlimVirgin: I take your point, Matt, but I feel you ought to have discussed this with the blocking admin, rather than undoing the block. . . .

This interaction prompted Jimmy "Jimbo" Wales, Wikipedia cofounder and leader, to write: "SlimVirgin, MattCrypto: this is why I love Wikipedians so much. I love this kind of discussion. Assume good faith, careful reasoning, a discussion which doesn't involve personal attacks of any kind, a disagreement with a positive exploration of the deeper issues."[5] Whereas Godwin's Law recognizes the tendency to think the worst of others, Wikipedia culture encourages contributors to treat and think of others well. For example, participants are supposed to abide by the norm of "Wikiquette," which includes the guidelines of "Assume Good Faith" (AGF) and "Please Do Not Bite the Newcomers."[6] Such Wikipedia norms and their relationship to the technology, discourse, and vision of a universal encyclopedia prompt me to ask: How should we understand this community's collaborative—"good faith"—culture? In the following chapters I offer my understanding on this question, but, first, an introduction.

Wikipedia

When I speak of Wikipedia I am referring to a wiki project, which includes both the textual artifact and the community producing it. (This is a common usage, as is referring to Web sites without a definite article, that is "I searched Google and Wikipedia" not "I searched the Google and the Wikipedia.") Furthermore, there is a particular vision of access and openness at Wikipedia, as seen in its slogan as "the free encyclopedia that anyone can edit." This vision, the encyclopedia, and its community and culture are introduced in the sections that follow.

The Vision

The Wikimedia Foundation, the nonprofit organization under which Wikipedia and its related projects operate, asks the reader to "Imagine a world in which every single human being can freely share in the sum of all knowledge. That's our commitment."[7] However, this commitment is not unique to the new millennium. Indeed, Wikipedia's heritage can be traced back to the beginning of the twentieth century. In particular, it can be traced back to Paul Otlet's Universal Repertory and H. G. Wells's proposal for a *World Brain* (included in a 1937 book of the same title). These projects were conceived as furthering increased access to information; facilitated by the (then relatively novel) technologies of the index card, loose-leaf binder, and microfilm. However, this vision exceeds the production of information. Wells proposed that reference work compilers would be joined by world scholars and international technocrats to produce a resource that every student might easily access, in a personal, inexpensive, and portable format. Furthermore, this collection of the world's intellect was envisioned to yield a greater sense of unity: Wells hoped that such an encyclopedia could solve the "jig-saw puzzle" of global problems by bringing all the "mental wealth of our world into something like a common understanding"; this would be more than an educational resource, it would be an institution of global mediation.[8] Wikipedia shares this concern for "the sum of all knowledge" with early visionaries. And while no one argues that Wikipedia will bring about world peace, I do argue goodwill is necessary to its production and an occasional consequence of participation.

However, while most early Wikipedians were probably unaware of these predecessors from a century ago there was a more immediate inspiration: Free and Open Source Software (FOSS). One of the earliest news articles about Nupedia, Wikipedia's non-wiki progenitor, notes: "The philosophy of the open-source movement is spreading within the industry. Now, a maker of a Web-based encyclopedia wants to apply its principles to share knowledge in general."[9] Nupedia described itself as "the open content encyclopedia" and was available under the GNU Free Documentation License. (GNU is a seminal free software project.) FOSS is licensed to enable users to read and improve on the source of the software they use. This has proven to be a much noted alternative to proprietary software in which one's usage can be restricted (e.g., unable to backup, install multiple copies, repair, or improve). When I emailed Jimmy Wales to ask about the influence of FOSS on his thinking, he replied:

> In general what I can say is that Nupedia was absolutely inspired by the free software movement. I spent a lot of time thinking about online communities and collaboration, and one of the things that I noticed is that in the humanities, a lot of people were collaborating in _discussions_, while in programming, something different was going on. People weren't just talking about programming, they were working together to build things of value.[10]

Consequently, the inspiration for a free and open source encyclopedia—in terms of access, cost, and collaboration—might be thought of as the most recent stage of a long-running pursuit.

The Encyclopedia

Wikipedia is the wiki-based successor to Nupedia and its name is a portmanteau of "wiki," an online collaborative editing tool, and "encyclopedia," itself a contraction of the Greek *enkyklios* and *paidei,* referring to the "circle of learning" of the classical liberal arts. This name is evidence of a geeky sort of linguistic humor and also prompts the question of whether a relatively open-to-all wiki can also be a high-quality reference work. In the following pages I return to these points but for now let's consider the wiki and encyclopedic aspects of the thing we call Wikipedia.

Wikipedia is an online wiki-based encyclopedia. "Wiki wiki" means "super fast" in the Hawaiian language, and Ward Cunningham chose the name for his collaborative WikiWikiWeb software in 1995 to indicate the ease with which one could edit pages. (He learned of the word during his first visit to Hawaii when he was initially confused by the direction to take the "Wiki Wiki Bus," the Honolulu airport shuttle.[11]) In a sense, the term *wiki* captures the original conception of the World Wide Web as both a browsing and editing medium; the latter capability was largely forgotten when the Web began its precipitous growth and the most popular clients did not provide their users with the ability to edit Web pages.

The wiki changed this asymmetry by placing the editing functionality on the server. Consequently, if a page can be read, it can be edited in any browser. With a wiki, the user enters a simplified markup into a form on a Web page. Using the Wikipedia syntax one simply types "# this provides a link to [[Ward Cunningham]]" to add a numbered list item with a link to the "Ward Cunningham" article. The server-side Wikipedia software translates this into the appropriate HTML and hypertext links. To create a new page, one simply creates a link to it, which remains red until someone actually adds content to its target destination. These capabilities are central to and representative of wikis.

Wikipedia now has a number of features that not all wikis share (although Wikipedia's open source MediaWiki platform is used by many other projects). Each wiki page includes links through which one can log in (if desired), bookmark (and "watch") the pages one cares about, or discuss how the page is being edited on its "Talk" or "Discussion" page—and this too is wiki. A history of a page is also available, showing all changes to the page (including the author, time, and edit summary); different versions can easily be compared. Two widely used features of Wikipedia are categories and templates. Users have the ability to label pages with categories, which are then used to automatically generate indexes. For example, the "1122 births" category page lists six biographical articles for those born in 1122.[12]

A wiki template is "a page which can be inserted into another page via a process called transclusion." These small template "pages" (usually no more than a few lines of text) "are used to add recurring messages to pages in a consistent way, to add boilerplate messages, to create navigational boxes and to provide cross-language portability of texts."[13] Templates are included on a page by including the template name within a pair of curly parentheses. So, with the inclusion of the "{{pp-vandalism}}" markup, a Wikipedia page will display a warning box that "this page is currently protected from editing to deal with vandalism." Many templates, such as the vandalism one, also add a category, creating an index of all pages presently using that template.[14]

The application of the wiki platform with a few encyclopedic features enables surprisingly sophisticated content creation.[15] And, as we will see throughout this book, wikis often are thought of as potent collaborative tools because they permit asynchronous, incremental, and transparent contributions from many individuals. Yet, as is often the case, the consequence of this quick and informal approach of editing the Web was not foreseen—or, rather, was pleasantly surprising. Wikipedia is the populist offshoot of Nupedia, started in March 2000 by Jimmy Wales and Larry Sanger. Nupedia's mission was to create a free encyclopedia via rigorous expert review under a free documentation license. Unfortunately, this process moved rather slowly and, having recently been introduced to wikis, Sanger persuaded Wales to set up a scratchpad for potential Nupedia content where anyone could contribute. However, there was "considerable resistance on the part of Nupedia's editors and reviewers to the idea of associating Nupedia with a wiki-style website. Sanger suggested giving the new project its own name, Wikipedia, and Wikipedia was soon launched on its own domain, wikipedia.com, on 15 January 2001."[16]

Wikipedia proved to be so successful that when the server hosting Nupedia crashed in September 2003 it was never restored. In August 2009 there were over "75,000 active contributors working on more than 10,000,000 articles in more than 260 languages"; the original English version includes more than three million articles, having long ago subsumed most of the original Nupedia content. Twenty-five other language editions have more than 100,000 articles.[17] These editions are evidence of the international character of the universal vision. (Within two weeks of the launch of Nupedia, Sanger wrote that he had already received offers to translate articles and that supporting this work should be a priority despite any delays it might introduce.[18]) The Wikimedia Foundation, incorporated in 2003, is now the steward of Wikipedia as well as a wiki-based dictionary, a compendium of quotations, a source of collaborative textbooks, a repository of free source texts, and a collection of images that can be used by other Wikimedia projects.

Given its size, it is no trivial task to understand what Wikipedia actually includes. (Claims by some to have read every word of an encyclopedia are impossible with Wikipedia.) However, recent research suggests that Wikipedia's topical coverage of general knowledge and technical issues is quite good, but it has blind spots in other specialist areas. Users seemingly are most interested in people, as articles about humans (e.g., biographies, culture, entertainment, the self, and sexuality) are the largest categories of content and the most visited by readers.[19]

Of course those three million articles on the English-language Wikipedia—the focus of this book—are not of equal quality. A summary from a proposal to achieve "100,000 Feature-Quality Articles," Wikipedia's highest quality level, reports that as of February 2009, roughly half of the English articles have been assessed for quality, and of those roughly 11,000 were considered to be of "Featured" (outstanding and thorough), "A" (very useful and fairly complete), or "Good" (useful to most readers with no obvious problems) quality.[20] External assessments of Wikipedia quality indicate it is at parity with general-purpose print reference works. In December 2005 the prestigious science journal *Nature* reported the findings of a commissioned study in which subject experts reviewed forty-two articles in Wikipedia and *Encyclopaedia Britannica*; it concluded "the average science entry in Wikipedia contained around four inaccuracies; Britannica, about three."[21] (This was widely reported, discussed, and eventually contested by *Britannica*.) In his 2007 study, George Bragues summarized preceding research by noting the balance "leans in favor of Wikipedia, though both the number of studies, and the proportion of positive to negative points, is far from enough to establish any firm conclusions." His comparison of the biographies of seven prominent philosophers across authoritative reference works led him to conclude that while Wikipedia sometimes failed to be consistent in topical coverage, "The sins of Wikipedia are more of omission than commission."[22] Clearly, there is much work to be done to make Wikipedia a consistently high-quality reference work. On the other hand, the breadth of so many articles means that Wikipedia has extraordinary and up-to-date coverage of even the narrowest interests. While it has yet to be assessed, and would fall short of featured article status, the article on my Brooklyn neighborhood, adjacent to Gowanus Canal, is quite handy.[23]

The Community

It can be even more difficult to get a sense of the English Wikipedia community. The "Editing Frequency" page indicates that 41,393 registered (logged in) users made five or more edits in September 2008 (the most recent published figures).[24] Yet this doesn't represent the hundreds of thousands of contributors who may have previously contributed, those who edit "anonymously" or without a consistent identity, and the "Wiki gnomes" who may fall short of this threshold—such as myself—but continue to quietly enact small tasks (e.g., fixing a typo) as opportunity presents itself and time permits.

Given Wikipedia's reputation as the encyclopedia "anyone can edit," an early research question was who did most of the work: the few who contribute a lot (i.e., the "elite") or the many who contribute a little (i.e., the "bourgeoisie," "long tail," "crowd," or "mob"). Jimmy Wales, Wikipedia cofounder, noted in December 2005 that "half the edits by logged in users belong to just 2.5% of logged in users." This conclusion has since been challenged, and the question itself has been reconsidered given that one's definition of "contribution" affects the results (e.g., authoring new content, editing, organizing, or discussing policy). Also, the nature of Wikipedia contribution may be changing as it matures. For example, one 2007 study showed that overall "conflict and coordination tasks" are growing faster than that of editing encyclopedic articles.[25] Even so, it appears there are significant differences in the amount and type of work Wikipedians do, and these proportions are changing, but that all contributions are important—if not synergistic.[26]

For those involved in administrative functions (e.g., protecting pages from vandalism), "There are 1,675 (as of now) administrator accounts (active and otherwise), 906 of them active (as of 2009-08-11)."[27] There are over two hundred names on the #wikipedia chat channel and there are more than seven hundred subscribers to the community *Wikizine* bulletin.[28] On the relatively high-traffic wikiEN-l list I counted approximately 180 unique posters in the first half of 2009, though I am confident this is a fraction of those subscribed. More topically, "Wikipedia Projects" (or "WikiProjects") are wiki pages in which contributors interested in a particular topic can plan and discuss their efforts; there are hundreds of active projects spanning the range of topics at Wikipedia.[29] (As evidence that the English Wikipedia is stabilizing, none of these numbers have changed significantly over the past year.) Again, because of the large and often pseudonymous character of contribution, it can be difficult to make any sort of demographic claims about Wikipedians. However, in April 2009 a preliminary report of a Wikipedian survey indicates respondents were predominantly young males; almost half have achieved an undergraduate degree[30]—but given their youth more may eventually do so.

I characterize my approach to my subject as a historically informed ethnography: observing—and occasionally participating in—the Wikipedia online community. While I make use of a broader historical context and online archives, I began to follow this community "in real time" in 2004 via a number of venues. First, there are the actual Wikipedia pages and edits to them; this includes the encyclopedic articles (e.g., "Chemistry") as well as the "meta" pages documenting the policies and norms of Wikipedia itself (e.g., "Neutral Point of View"). Second, there is the talk/discussion page associated with each article on which conversation about the article occurs (e.g., suggestions for improvements). Third, there are mailing lists on which more abstract or particularly difficult issues are often discussed; wikiEN-l and wiki-l often include discussions of the administration and policies of Wikipedia. Also, there are the *Wikipedia Signpost* and *Wikizine* newsletters, other community forums such as the popular "Village Pump," and various Wikipedia-related blogs, aggregators, and podcasts.[31] Fifth, and finally, there are the physical spaces in which some community members interact. Through Wikipedia "meetups" I've attended in New York and annual Wikimania conferences I've met a couple dozen contributors. It's quite easy to speak to a new Wikipedian acquaintance about issues of concern to the community, and many of these people I've spoken to more than once. These conversations were informative, but casual. I formally interviewed only a handful of sources and otherwise have relied on the public activity and discourse of the community.

In sum, there are the tens of thousands of active contributors who are familiar with the basic practices and norms of English Wikipedia. This includes smaller communities on the scale of hundreds or dozens of members within geographical, functional, and topical boundaries. And the English Wikipedia is part of a larger community of multilingual encyclopedias and Wikimedia projects.

The Culture

The focus of this book is Wikipedia's collaborative culture. While I explain what I mean by this in chapter 3, I want to first briefly introduce my approach and Wikipedia's core collaborative principles.

In addition to millions of encyclopedic articles, Wikipedia is suffused with a coexisting web of practices, discussion, and policy pages, the latter of which populate the Wikipedia "project namespace" and "Meta-Wiki" of Wikimedia projects.[32] A charming example of wiki practice is the awarding of a "barnstar," an image placed on another's user page to recognize merit. "These awards are part of the Kindness Campaign and are meant to promote civility and WikiLove. They are a form of warm fuzzy: they are free to give and they bring joy to the recipient."[33] There are different stars for dozens of virtues, including random acts of kindness, diligence, anti-vandalism, good humor, resilience, brilliance, and teamwork. As in any other community, at Wikipedia there is also a history of events, set of norms, constellation of values, and common lingo. Also, not surprisingly, there is a particular sensibility, including a love of knowledge and a geeky sense of humor. Unlike many other communities, most all of this is captured online. Even beyond the inherent textual verbosity of other online communities, Wikipedia is extraordinarily self-reflective. Most everything is put on a wiki, versioned, linked to, referenced, and discussed. And in the tradition of Godwin's Law of Nazi Analogies, an initial set of four observations by Wikipedian Raul654 in 2004 has become a collection of over two hundred laws by Wikipedians describing their own interactions. This proliferation is itself the subject of Norbert's Law: "Once the number of laws in a list exceeds a critical mass (about six), the probability of new laws being tortured, unfunny and bland rises rapidly to unity."[34] Furthermore, the "WikiSpeak" essay is an ironic glossary of terms that gives insight into both Wikipedia's substance and faults. For example, collaboration is defined as "One editor taking credit for someone else's work."[35] ("Raul's Laws of Wikipedia" and "WikiSpeak" definitions also make appearances as chapter epigraphs in this book.)

This wealth of material is a treasure given my interest in understanding how people make sense of their experiences of working together. And while

I was influenced by varied scholars in conceiving of and executing this work, sociologist Harold Garfinkel's "ethnomethodology" is particularly relevant given its focus on "practical activities, practical circumstances, and practical sociological reasoning."[36] By "practical sociological reasoning" Garfinkel means the discourse and reasoning of the actual participants themselves. How a community makes sense of its experience is what Garfinkel refers to as "accounting processes." As Alain Coulon writes in the introduction to *Ethnomethodology*, it is "the study of the methods that members use in their daily lives that enable them to live together and to govern their social relationships, whether conflictual or harmonious"; that is, how "the actor undertakes to understand his action as well as that of others."[37] The two hundred-plus laws posited by Wikipedians are a salient example of the community trying to understand itself and its circumstances.

Therefore, much of this book is an exploration of the norms guiding Wikipedia collaboration and their related "accounting processes," but there are three core policies central to understanding Wikipedia and are worthwhile addressing at the outset: "Neutral Point Of View" (NPOV), "No Original Research," and "Verifiability." While NPOV at first seems like an impossible, or even naïve, reach toward an objectively neutral knowledge, it is quite the opposite. The NPOV policy instead recognizes the multitude of viewpoints and provides an epistemic stance in which they all can be recognized as instances of human knowledge—right or wrong. The NPOV policy seeks to achieve the "fair" presentation of all sides of the dispute.[38] Hence, the clear goal of providing an encyclopedia of all human knowledge explicitly avoids many entanglements. Yet when disagreements do occur they often involve alleged violations of NPOV. Accusations of and discussions about bias are common within the community and any "POV pushing"—as Wikipedians say—is seen as compromising the quality of the articles and the ability for disparate people to work together. However, violations of NPOV are not necessarily purposeful, but can result from the ignorance of a new participant or the heat of an argument. In some circumstances, the debate legitimately raises substantive questions about NPOV. In any case, while some perceive NPOV as a source of conflict, it may act instead as a conduit: reducing conflict and otherwise channeling arguments in the productive context of developing an encyclopedia.

The other two policies of "No Original Research" and "Verifiability" are both about attribution, meaning, "All material in Wikipedia must be attributable to a reliable, published source."[39] The latter principle obviously is important for an encyclopedia that "anyone can edit." Furthermore, and perhaps more importantly, the notion of "No Original Research" permits the community to avoid arguments about pet theories and vanity links (i.e., when a person links from Wikipedia to a site they wish to promote). If you have "a great idea that you think should become part of the corpus of knowledge that is Wikipedia, the best approach is to publish your results in a good peer-reviewed journal, and then document your work in an appropriately nonpartisan manner."[40] Since Wikipedia does not publish original research, "Verifiability" then implies that "readers are able to check that material added to Wikipedia has already been published by a reliable source, not whether we think it is true."[41] In a sense, a Wikipedia article can be no better than its sources.

These three policies of "Neutral Point of View," "No Original Research," and "Verifiability" have been characterized as the "holy trinity" of Wikipedia,[42] without one being preeminent over any other, according to Wales:

> I consider all three of these to be different aspects of the same thing, ultimately. And at the moment, when I think about any examples of apparent tensions between the three, I think the right answer is to follow all three of them or else just leave it out of Wikipedia. We know, with some certainty, that all three of these will mean that Wikipedia will have less content than otherwise, and in some cases will prevent the addition of true statements. For example, a brilliant scientist conceives of a new theory which happens to be true, but so far unpublished. We will not cover it, we will not let this scientist publish it in Wikipedia. A loss, to be sure. But a much much bigger gain on average, since we are not qualified to evaluate such things, and we would otherwise be overwhelmed with abject nonsense from POV pushing lunatics. There is no simple a priori answer to every case, but good editorial judgment and the negotiation of reasonable people committed to quality is the best that humans have figured out so far.:) —Jimbo Wales 15:33, 15 August 2006 (UTC)[43]

Such Wikipedia policies and Wikipedian discourse are central to the cultural account I present. I approached these policies and discourse as a "naturalistic inquiry," that is, an emergent process in which I collected texts relevant to Wikipedia collaboration; these were categorized, carefully considered, and refined as I continued to engage the community.[44] As of August 2009 I've collected over 1,300 Wikipedia-related primary sources.[45] (I present such sources mostly verbatim—with minimal corrections or editorial caveats such as "[sic]"—and use names as provided, including idiosyncratic spellings.) I've also participated in the community by attending meetings and editing a few pages. While there is a temptation to alter Wikipedia to suit one's purposes—as journalists, lobbyists, and disputing Wikipedians have discovered—I've purposively avoided policy and edit disputes. As organizational ethnographer John Van Maanen writes, "an ethnography is a written representation of a culture" and so I've attempted to "display the culture in a way that is meaningful to readers without great distortion."[46] And much as Henry Jenkins admits that his fondness for the fan cultures he studies might "color" what is said, it also implies "a high degree of responsibility and accountability to the groups being discussed."[47]

This Book, in Short

A hazard in thinking about new phenomena—such as the Web, wiki, or Wikipedia—is to aggrandize novelty at the expense of the past. To minimize this inclination I remind myself of the proverb "the more things change, the more they stay the same." Therefore, I begin in chapter 2 with an argument that Wikipedia is an heir to a twentieth-century vision of universal access and goodwill; an idea advocated by H. G. Wells and Paul Otlet almost a century ago. This vision is inspired by technological innovation—microfilm and index cards then, digital networks today—and driven by the encyclopedic impulse to capture and index everything known. In some ways my argument is an extension of that made by historian Boyd Rayward who notes similarities between Paul Otlet's information "Repertory" and Project Xanadu, an early hypertext system.[48] My effort entails not only showing similarities in the aspirations and technical features of these older visions and Wikipedia, but also recovering and placing a number of Wikipedia's predecessors (e.g., Project Gutenberg, Interpedia, Nupedia) within this history.

In chapter 3, I turn to an essential feature linking Wikipedia to the pursuit of the universal encyclopedia and to Wikipedia's success: its *good faith collaborative culture*. While the relevance of "prosocial" norms has been noted by other scholars (along with notions of trust, empathy, and reciprocity), Wikipedia provides an excellent opportunity, because of its reflective documentation and discourse, to see how such norms emerge and how they are enacted and understood. I focus on the norms of "Neutral Point of View" and "Assume Good Faith" to argue that an open perspective on both *knowledge claims* and *other contributors*, respectively, makes for extraordinary collaborative potential. However, unlike the incompletely realized potential of earlier visions, Wikipedia is very real and very messy. How the community wrestles with issues of openness, decision making, and leadership can offer insight into collaborative cultures.

A facet of the universal encyclopedic vision has been an increase in the accessibility of knowledge. Wikipedia takes this further, by increasing access to information and its production. In chapter 4 I present the Wikipedia community as an *open content community*. This notion is inspired by FOSS and the subsequent popularization of "openness," but focuses on community rather than copyright licenses. I then consider four cases that challenge Wikipedia's openness as "the free encyclopedia that anyone can edit." In the first case I ask: is Wikipedia really something anyone can edit? That is, when Wikipedia implemented new technical features to help limit vandalization of the site, did it make Wikipedia more or less open? In the second case I describe the way in which a maturing open content community's requirement to interact with the sometimes "closed" world of law affects its openness. In this case, I review Wikipedia's "office action" in which agents of Wikipedia act privately so as to mitigate potential legal problems, though this is contrary to the community values of deliberation and transparency. Third, I briefly review concerns of how bureaucratization within Wikipedia itself might threaten openness. Finally, I explore a case in which a closed (female-only) group is set up outside of, and perhaps because of, the "openness" of the larger Wikipedia community.

Beyond the more abstract question of openness, the fact that the community has a porous boundary and a continuous churn of pseudonymous and anonymous users means there are significant challenges in working together and making decisions. H. G. Wells thought his "World Encyclopedia" should be more than an information repository; it should also be a "clearinghouse of misunderstandings."[49] By reviewing a specific "misunderstanding" about the naming of television show articles, I explore the benefits, challenges, and meaning of consensus at Wikipedia. Specifically, by contextualizing Wikipedia practice relative to other communities (e.g., Quakers and Internet standards organizations) I show how consensus is understood and practiced despite difficulties arising from the relative lack of resources other consensus communities have from the start.

And just as the complexities inherent in the understanding and practice of good faith, openness, and consensus reveal the character of Wikipedia and prompt insights into human interaction, the question of leadership in this type of community is also revealing. In open content communities, like Wikipedia, there often is a seemingly paradoxical use of the title "benevolent dictator" for leaders. In chapter 6, I explore discourse around the use of this moniker so as to address how leadership works in open content communities and why Wikipedia collaboration looks the way it does today. To do this, I make use of the notion of "authorial" leadership: leaders must parlay merit resulting from authoring something significant into a form of authority that can also be used in an autocratic fashion, to arbitrate between those of good faith or defend against those of bad faith, with a soft touch and humor when—and only when—necessary.

Finally, in chapter 7 I focus on the cultural reception and interpretation of Wikipedia. The collaborative way in which Wikipedia is produced has caught the attention of the world. Discourse about the efficacy and legitimacy of such a work abound, from the news pages of the *New York Times* to the satire of *The Onion*. Building on the literature concerning the controversies that have surrounded other reference works, such as Harvey Einbinder's *The Myth of the Britannica* and Herbert Morton's *The Story of Webster's Third*,[50] I make a broader argument that reference works can serve as a flashpoint for larger social anxieties about technological and social change. With this understanding in hand, I try to make sense of the social unease embodied in and prompted by Wikipedia by way of four themes: collaborative practice, universal vision, encyclopedic impulse, and technological inspiration.

I conclude with a reflection upon H. G. Wells's complaint of the puzzle of wasted knowledge and global discord. Seventy years later, Wikipedia's logo is that of a not yet complete global jigsaw puzzle. This coincidence is representative of a shared dream across the decades. The metaphor of the puzzle is useful in understanding Wikipedia collaboration: NPOV ensures that we can join the scattered pieces of what we think we know and good faith facilitates the actual practice of fitting them together.

2 The Pursuit of the Universal Encyclopedia

sum of all human knowledge *adj.*: Descriptive of the ultimate aim of Wikipedia. Regretfully, the vast majority of human knowledge is not, in actual fact, of interest to anyone, and the benefit of recording this collective total is dubious at best.
—WikiSpeak

In March 2000 Jimmy Wales, cofounder of Wikipedia and its Nupedia progenitor, sent his first message to the Nupedia email list: "My dream is that someday this encyclopedia will be available for just the cost of printing to schoolhouses across the world, including '3rd world' countries that won't be able to afford widespread internet access for years. How many African villages can afford a set of Britannicas? I suppose not many. . . ."[1] In this statement one can find a particular type of Enlightenment aspiration: a *universal* encyclopedic vision of increased information access and goodwill. Richard Schwab, a scholar of the Enlightenment and *Encyclopédie,* wrote that at that time thinkers were coming to recognize that "cumulatively they were a force in the world" and possessed "a new solidarity and power to advance human affairs."[2] For example, Denis Diderot (1713–1784), editor of the *Encyclopédie,* wrote that it was developed by a society of men working separately on their respective tasks "but all bound together solely by their zeal for the best interests of the human race and the feeling of mutual good will." The aim of this effort was "to collect all the knowledge that now lies scattered over the face of the earth, to make known its general structure to men among whom we live, and to transmit it to those who will come after us."[3] Historian Richard Yeo describes the universalistic principles underlying so much of the Enlightenment culture as "the ideal of transportable knowledge, the communication of ideas across national and confessional boundaries; the ability of individuals, where ever they lived, whatever their social status, to participate in a universal conversation. . . . These [reference] works offered the possibility of a reliable codification of knowledge by seeking to record any consensus, and by fixing the meaning of terms." As an example of unbridled optimism, Yeo cites the French social thinker and encyclopedist Henri Saint-Simon's (1760–1825) forecast that lasting peace could be achieved between France and England within a year of jointly undertaking work on a "New Encyclopedia."[4]

Granted, one can poke fun at the pretense of summarizing "all human knowledge," as is done in the WikiSpeak definition at the start of this chapter. Creating a world encyclopedia, much less world peace, is a difficult task and the trivia found on Wikipedia is a source of delight to some and derision to others. Nonetheless, the coupling of increased information access with human accord is a long-held dream. While its advocates are sometimes overly exuberant they can also be pragmatic, as seen in Wales's 2004 "Letter from the Founder":

> Our mission is to give freely the sum of the world's knowledge to every single person on the planet in the language of their choice, under a free license, so that they can modify, adapt, reuse, or redistribute it, at will. And, by "every single person on the planet," I mean exactly that, so we have to remember that much of our target audience is not yet able to access the Internet reliably, if at all. . . . Our community already comes from a huge variety of backgrounds, and over time the variety will only increase. The only way we can coordinate our efforts in an efficient manner to achieve the goals we have set for ourselves, is to love our work and to love each other, even when we disagree. Mutual respect and a reasonable approach to disagreement are essential . . . on this incredible ridiculous crazy fun project to change the world.
>
> None of us is perfect in these matters; such is the human condition. But each of us can try each day, in our editing, in our mailing list posts, in our irc [Internet Relay] chats, and in our private emails, to reach for a higher standard than the Internet usually encourages, a standard of rational benevolence and love.[5]

Furthermore, whereas Wales conceived of his encyclopedia reaching those without access to the Internet, technology is central to the modern version of the vision.[6] Technology is expected to facilitate a radically accessible resource that bridges the distance between people. As recounted in Tom Standage's history of the telegraph (i.e., the "Victorian Internet"), the "rapid distribution of news was thought to promote universal peace, truthfulness, and mutual understanding."[7] H. G. Wells felt that "Encyclopaedic enterprise has not kept pace with material progress" but when the "modern facilities of transport, radio, [and] photographic reproduction" were embraced the creation of a permanent world encyclopedia would be "a way to world peace": "Quietly and sanely this new encyclopaedia will, not so much overcome these archaic discords, as deprive them, steadily but imperceptibly, of their present reality."[8] One can even see the universal vision in a different sort of technology altogether: the airplane. Joseph Corn, in *The Winged Gospel*, tells of high aeronautical expectations. "Air Globes," representations of the earth and its cities without political or geographical boundaries, were deployed in the classroom to tangibly symbolize "the new world which Americans believed the airplane was about to create, a world of peace where national boundaries and topographical features were no longer pertinent."[9]

This technological inspiration and aspiration for global accord is quite in keeping with the heritage of Wikipedia described in this chapter—and globes are a recurrent motif in this work. And while one can draw parallels between Wikipedia, the Enlightenment, and even aeronautics or ancient encyclopedic efforts, I focus on the twentieth century. (Contemporary visionaries frequently reference the ancient world's lost Library of Alexandria as a historic predecessor; the desire to link the vision across the ages is also seen in the announcement of the new Bibliotheca Alexandria hosting the Wikimania 2008 conference.[10]) In the following pages I touch upon important moments in the pursuit of the universal encyclopedia, a technologically inspired reference work with progressive intentions, from early documentalists to Wikipedia; along the way I also ask why it took so long for this vision to become a reality.

The Index Card and Microfilm

The idea of a personal encyclopedic device is frequently attributed to Vannevar Bush (1890–1974), an electrical engineer and advocate of America's war research program. In a 1945 article entitled "As We May Think," he famously outlined the idea for a *memex*, an "enlarged intimate supplement" to memory. This was envisioned as an electromechanical microfilm device "in which an individual stores all his books, records, and communications, and which is mechanized so that it may be consulted with exceeding speed and flexibility." Noting that for a nickel, the *Britannica* could be placed on 8.5-by-11-inch microfilm and mailed anywhere for a cent, Bush predicted "wholly new forms of encyclopedias will appear, ready-made with a mesh of associative trails running through them, ready to be dropped into the memex and there amplified." Thus science might continue to advance without forgetting the wisdom of hard-earned experience, including science's darker, "cruel," applications in warfare.[11]

However, the memex was proposed in a larger context of lesser-known microfilm technologies and innovators. In fact, the *beginning* of the twentieth century was a seminal period for internationalism and information science. In America, Melvil Dewey (1851–1931) proffered his decimal system, founded library institutions, attempted American spelling reform, and advocated for the metric system. In Germany, Wilhelm Ostwald (1853–1932) invested much of the award from his Nobel Prize for Chemistry in Die Brücke ("The Bridge"), an international institute he cofounded for organizing intellectual work across the world. Emanuel Goldberg (1881–1970), once a student of Ostwald's, advanced imaging technologies and developed a "Statistical Machine" by which microfilm records could be indexed and subsequently retrieved. In France, Suzanne Briet (1894–1989) pioneered information services at the Bibliothèque Nationale and internationally advanced the theoretical, educational, and institutional development of documentation.[12] Such efforts were often inspired by a new awareness of global interdependence and contemporary technologies such as the index card and microfilm. Also, early documentalists were of a time—one might even say an "age" or "movement"—when the efforts and ideas of one inspired and influenced others. One can trace Wikipedia's heritage back to this period, as seen most clearly in the writings of two twentieth-century visionaries: Paul Otlet and H. G. Wells.

Paul Otlet and the Universal Bibliographic Repertory

As a boy, the Belgian Paul Otlet (1868–1944) played at the task of extracting and organizing knowledge: he and his brother drew up a charter for a "Limited Company of Useful Knowledge."[13] At the age of eighteen he wrote in his diary, "I write down everything that goes through my mind, but none of it has a sequel. At the moment there is only one thing I must do! That is, to gather together my material of all kinds, and connect in with everything else I had done up till now."[14] Throughout his life Otlet was beholden to a vision of technology as a means of dissembling, synthesizing, and distributing knowledge on an international scale. Later in life, in 1918, he wrote his vision was supported by "three great trends" of his time: "the power of associations, technological progress and the democratic orientation of institutions."[15] New technologies of the day included loose-leaf binders, index cards, and microphotography. For example, Otlet and Robert Goldschmidt, an engineer and microphotography pioneer, estimated that a small can of film could hold 80,000 square meters of photographic documents such that books would soon be compact, light, permanent, inexpensive, durable, and easy to produce, conserve, and consult.[16] However, it was the humble 3-by-5-inch index card that would be the basis of some of his most profound insights.[17]

Otlet's career began in earnest in 1891 when he joined, as a young lawyer, the Society for Social and Political Studies, under Henri La Fontaine, director of the Society's bibliographic program and future collaborator. The span of Otlet's career would include helping found a universal bibliographic database/encyclopedia, an international library and museum, and numerous international associations.[18] Furthermore, in his most famous publication of 1934, *Traité de Documentation,* he wrote of a desk in the form of a wheel from which different projects (workspaces) could be switched as they rotated—foreshadowing the multiple desktops and tabs of contemporary computer interfaces. Inspired by the arrival of radio, phonograph, cinema, and television, Otlet also posited that there were as yet many "inventions to be discovered," including the reading and annotation of remote documents and computer speech.[19]

Yet there are three less speculative ideas that I think are particularly relevant to today's Web and Wikipedia: the Repertory, the Universal Decimal Classification (UDC), and the "monographic principle." The first two projects, of information collection and classification, arose following the first International Conference of Bibliography, in 1895, and were carried out at the newly created International Institute of Bibliography (IIB).[20] What began as a bibliographic "Repertory" of author files and classified subject files came to include repertories of image files ("iconographic") in 1905 and full text files ("dossiers") in 1907. The collections grew quickly; by 1912 the bibliographic repertory contained over nine million entries, and by 1914 the textual dossier "contained a million items in 10,000 subject files."[21] Much like the millions of Wikipedia articles, and its thousands of categories and lists, we see a shared concern with collecting and ordering the world's knowledge.

Of course, how could one possibly refer to and access all of this information? Otlet proposed a classification scheme which anticipated more recent information technologies. The UDC, based on the Dewey Decimal System, spanned over two thousand pages in its first full edition (1904–1907). Much like the URLs of today's Web, the UDC enabled one to refer to materials deposited in the inventory of the Repertory. In addition to the relatively simple—though extensive—scheme of decimal division, Otlet complemented his system with a set of symbols specifying addition, extension, algebraic subgrouping, and language.[22] For example the UDC notation "069.9(100)'1851'(410.11)" specifies The Great Exhibition of 1851 in London.[23] This capability made the UDC more than a classification system: it was a primitive query language permitting one to specify a subset of the catalog. Indeed, a search service with documented guidelines for queries was provided until the early 1970s.[24] However, while being able to uniquely identify a resource on the Web today continues to be important, the way in which we manage most online information is no longer so carefully organized—and this is thought to be a feature to some, rather than a bug. As David Weinberger writes in his book *Everything Is Miscellaneous: The Power of the New Digital Disorder,* we might think of attempts at managing information in three historical orders: the order of manipulation and ordering in which physical objects are arranged (e.g., the shelving of a book); the use of metadata and the catalog (i.e., using a second-order surrogate—where we might place Otlet's innovations); and today's disorder (i.e., fluid, ad hoc, temporary, and disposable). Weinberger writes, "The third order takes the territory subjugated by classification and liberates it. Instead of forcing it into categories, it tags it."[25] As I note in chapter 7, the notion of the "crowd" creating and managing knowledge is profoundly disturbing to some, as is a notion I think even more relevant to discourse about knowledge: the "monographic principle."[26]

This primary tenet of Otlet's schemes permitted one to "detach what the book amalgamates, to reduce all that is complex to its elements and to devote a page [or index card] to each." Pages and cards would not be bound, but "movable, that is to say, at any moment the cards held fast by a pin or a connecting rod or any other method of conjunction can be released."[27] (This is a perfect example of the ideal of "transportable knowledge" Richard Yeo speaks of with respect to the Enlightenment.) Elsewhere, Otlet wrote, "The external make-up of a book, its format, the personality of its author are unimportant provided that its substance, its sources and its conclusions are preserved and can be made an integral part of the organization of knowledge."[28] Otlet envisioned being able to condense books and to strip them of their opinion so as to create "good" abstracts and even "scientific" book reviews.[29] As he wrote of his vision in 1903, the result, "This book, the 'Biblion,' the Source, the permanent Encyclopedia, the Summa" would "constitute a systematic, complete current registration of all the facts relating to a particular branch of knowledge. It will be formed by linking together materials and elements scattered in all relevant publications."[30] This sounds much like today's Web and Wikipedia. Furthermore, Wikipedia's concern with attribution (i.e., the "Verifiability" and "No Original Research" policies) can be seen in an intention of Otlet for his own project: "Readers, abstractors, systematisers, abbreviators, summarizers and ultimately synthesizers, they will be persons whose function is not original research or the development of new knowledge or even teaching existing systematic knowledge. Rather their function will be to preserve what has been discovered, to gather in our intellectual harvests, [to] classify the elements of knowledge."[31]

Additionally, beyond his dedication to the technological component of the universal vision, Otlet was an internationalist and supported the foundation of the League of Nations and The International Institute for International Cooperation (which would become UNESCO) with Henri La Fontaine. Whereas La Fontaine would be recognized with a 1913 Noble Peace Prize for his international efforts, Otlet's documentation efforts were largely forgotten. Prior to World War II the Belgian government withdrew its funding, and many of the holdings were lost when the Repertory's home (at the Palais du Cinquantenaire) was occupied by German forces; over subsequent decades the collections fell into disuse and obscurity.

While the UDC is perhaps Otlet's most lasting contribution, it was not until historian Boyd Rayward "rediscovered" Otlet and argued that his vision anticipated the early hypertext of Ted Nelson's Project Xanadu that Otlet was again appreciated by those interested in the history of information science.[32] (And interest has become even more wide ranging: in June 2008 the *New York Times* published an article about Otlet and the establishment of a new museum and archive in Belgium.[33]) Rayward wrote his essay "Visions of Xanadu" in 1994, unaware of the nascent Web and that Xanadu would never be deployed. In this chapter I make a similar argument except that I believe Otlet's Repertory foreshadowed Wikipedia. The Repertory was international, multilingual, collaborative, and predicated on technological possibility, much like Wikipedia today.

H. G. Wells and the "World Brain"

H. G. Wells, the English novelist famous for his science fiction, was also captivated by advances in technology and the notion of a universal reference work. (While Rayward could find no evidence of direct contact between Wells and Otlet, he thinks it very likely that they at least knew of each other from their mutual attendance at the 1937 Documentation Congress in Paris.[34]) Like Otlet, Wells's notion of a universal reference work was not an immediate and solitary brainstorm; it was the culmination of a number of long-standing interests as prompted by the circumstances of his time. First, in his outline of a *Modern Utopia* in 1905, Wells wrote of the implications of index cards:

> A little army of attendants would be at work upon this index day and night . . . constantly engaged in checking back thumb-marks and numbers, an incessant stream of information would come, of births, of deaths, of arrivals at inns, of applications to post-offices for letters, of tickets taken for long journeys, of criminal convictions, marriages, applications for public doles and the like. . . . So the inventory of the State would watch its every man and the wide world write its history as the fabric of its destiny flowed on.[35]

Second, since at least 1928, Wells had been advocating for an internationalist revolution, one world government, or "Open Conspiracy."[36] Information historian Dave Muddiman aptly identifies the key elements of this "modern" program as: "universalism and the 'World State'; planning and a central organization; a faith in scientific and technical advance, education, professionalism, expertise and benevolent socialism."[37] (Understandably, the idea of a "World State" that kept tabs on its citizens in such a pervasive manner can be thought to be more dystopian than otherwise and counter to our current political sentiments;[38] but Wells thought of these things in an optimistic light.)

Third, Wells was beginning to think of artifacts (like books) and institutions (like museums) as a type of "super-human memory" that would prompt a mental expansion for which "the only visible limit is our planet and the entire human species."[39] Each of these threads found their way into his 1936 proposal for a world encyclopedia, or, as he liked to call it, a "World Brain." (Otlet, too, at least once made reference to an "artificial brain" and Wilhelm Ostwald wrote of a "Gehirn der Welt" (World Brain) in 1912.[40]) Given advances in technology and the insecurity of the interwar period, Wells believed that intellectual resources were squandered: "We live in a world of unused and misapplied knowledge and skill" and "professional men of intelligence have great offerings but do not form a coherent body that can be brought to general affairs." He hoped that a world encyclopedia could "solve the problem of that jig-saw puzzle and bring all the scattered and ineffective mental wealth of our world into something like a common understanding." Beyond producing a resource for students and scholars, it would be an institution of "adjustment and adjudication; a clearinghouse of misunderstandings." Ultimately, he hoped that such an institution would further the movement toward "unification and perhaps the abandonment of war." But "without a World Encyclopedia to hold men's minds together in a common interpretation of reality, there's no hope whatever of anything but an accidental and transitory alleviation to any of our world troubles."[41]

With respect to technology, given the resources of "micro-photography" Wells felt: "the time is close at hand when a student, in any part of the world, will be able to sit with his projector in his own study at his or her convenience to examine any book, any document, in exact replica." And much like one of Wikipedia's greatest strengths, it need not limit itself as a "row of volumes printed and published once and for all" but could instead be "a sort of mental clearinghouse for the mind, a depot where knowledge and ideas are received, sorted, summarized, digested, clarified, and compared" in "continual correspondence" with all that was happening in the world.[42] Furthermore, he proposed that the encyclopedia be in a single language (English) as it was difficult to otherwise conceive of a polyglot project satisfying his goal of social unity. Yet, it is also difficult to conceive how any such project could be genuinely universal when limited to a single language. In the case of Wikipedia, it began as an English-language work and this version remains the largest, but there are now encyclopedias in other languages. While policy for the Wikimedia projects at large continues to be discussed on the English-language email lists and the Meta wiki, the different language communities are largely autonomous. (However, multilingual participants do often participate across different projects, and encyclopedic articles in one language can now link to their alternative language versions.) For Wells, perhaps English was only the initial, expedient step as he expected a universal (English-like) language would ultimately prevail in his "Modern Utopia."[43]

In keeping with the universal vision, and anticipating a key Wikipedia norm, H. G. Wells was concerned that his World Brain be an "encyclopedia appealing to all mankind," and therefore it must remain open to corrective criticism, be skeptical of myths (no matter how "venerated") and guard against "narrowing propaganda."[44] This strikes me as similar to the pluralism inherent in the Wikipedia "Neutral Point of View" goal of "representing significant views fairly, proportionately, and without bias."[45] And even going beyond Wikipedia's "No Original Research" policy, Wells thought the World Brain repository "should consist of selections, quotations, and abstracts as assembled by authorities—one need not create summaries."[46] However, one must note that his propensity to write with others' works close at hand, as he admits to with respect to the *Britannica*, perhaps led to plagiarism that was also indicative of larger character faults.[47]

In any case, despite claims of plagiarism, character faults, and his utopian expectations that people find alternatively progressive and frightening, Wells was a dedicated internationalist and forever looking toward the future. Like La Fontaine and Otlet, Wells thought the examples of The League of Nations, the Committee on Intellectual Cooperation in Paris, and the World Congress of Documentation were models, in spirit and application, for his own project. Yet unlike Otlet's efforts, which were well known in their time, the World Brain never materialized beyond the ardent vision of an author.

Digital Computers and Networks

I argue that in the first half of the twentieth century (via the examples of Otlet and Wells) we can discern a technologically inspired vision of a universal encyclopedia. This vision included collaborative capabilities—or, as Vannevar Bush spoke of, "amplifying" the contributions of others. For Otlet and Wells this collaboration was also part of their internationalist commitment. "Madame Documentation" Suzanne Briet captured this sentiment when she wrote of her library's reading room of three hundred patrons: "peaceful with their books. Peace through books."[48] Even Bush, an architect of the atomic weapons program, hoped a better (machine-aided) memory would not let us forget the horrors of war. Yet, in the first half of the twentieth century, these visions were never satisfactorily fulfilled. For one, microfilm wasn't up to the task. As Anthony West wrote in a biography of his father H. G. Wells, "he saw before too long the technology for the storage and retrieval systems that such a thing would require was still lacking, and that its time had not yet come."[49] However, in the latter half of the twentieth century a new technology, the computer network, engendered new possibilities and thus inspired new directions in the creation of encyclopedias. And while the expectation that a networked encyclopedia would herald a new era of world peace lessened, the likelihood of a widely accessible and collaborative encyclopedia increased.[50] Even so, why did it take so long for the vision of "wholly new forms of encyclopedias" to be realized in the form of Wikipedia? In this section I use moments from the history of hypertext and digital networks to argue that it required an alignment of a coherent goal, technical practicality, and serendipity: vision, pragmatics, and happenstance.

Project Xanadu

One ancestral line of Wikipedia's digital lineage is that of hypertext, including Ted Nelson's Project Xanadu, "the original hypertext and interactive multimedia system." Nelson's initial 1960 design "showed two screen windows connected by visible lines, pointing from parts of an object in one window to corresponding parts of an object in another window."[51] (Because Nelson coined the terms *hypertext* and *hypermedia*, Xanadu naturally is the first system to be labeled as such, though as Nelson himself notes, "Douglas Engelbart's NLS system at Stanford Research Institute was really the first hypertext system." Engelbart was the inventor of "word processing, outline processing, screen windows, the mouse, [and] the text link"[52]—and is discussed briefly in the next chapter.) Since then, Project Xanadu has had a complicated history of redesigns and attempts at commercial viability. Yet despite such difficulties, as conceived there were significant parallels between this work and its predecessors.[53] Its very name is a playful reference to the more fanciful aspirations of the universal vision. (The name was chosen in honor of Samuel Taylor Coleridge's unfinished poem "Kubla Khan."[54]) And the influence of Bush's memex on Nelson is evidenced by Bush's essay appearing as an appendix in Nelson's seminal *Literary Machines*.[55] Even the index card appears in journalist Kevin Kelly's story of meeting Nelson in 1984:

> Wearing a ballpoint pen on a string around his neck, he told me—way too earnestly for a bar at 4 o'clock in the afternoon— about his scheme for organizing all the knowledge of humanity. Salvation lay in cutting up 3 x 5 cards, of which he had plenty. . . . He spoke of "transclusion" and "intertwingularity" as he described the grand utopian benefits of his embedded structure. It was going to save the world from stupidity.[56]

This utopian characterization is not a journalist's fancy. In his book *Literary Machines* Nelson called for the support of "brilliant people looking for adventure and a challenge." He declared: "We have to save mankind from an almost certain and immediately approaching doom through the application, expansion and dissemination of intelligence. Not artificial, but the human kind."[57] This apocalyptic fear is echoed in H. G. Wells's parallel call "for a gigantic effort to pull together the mind of the race before it is altogether too late" as he feared that without "an educational revolution, a new Encyclopedism . . . we shall, as humanity, perish."[58]

However, as the Web gained popular attention, Nelson and his supporters came to feel the potential and priority of the vision he had been advocating for two decades were not satisfactorily respected. Indeed, an irony of the universal vision is that its proponents are often unfamiliar with their predecessors and disappointed with their successors.[59] So, when the Web began its precipitous growth, largely incognizant of Xanadu, Nelson and his colleagues developed a defensive attitude toward this upstart and their own portrayal in the press. A 1995 *Wired* article entitled "The Curse of Xanadu" prompted a particularly irked and detailed response from Nelson.[60] This contention between proponents of Xanadu and those of the Web has quieted to some extent in recent years because of the acknowledged dominance of the Web and the honors now accorded to Nelson. Even so, Project Xanadu's mission statement still reads: "The World Wide Web (another imitation of paper) trivializes our original hypertext model with one-way ever-breaking links and no management of version or contents. WE FIGHT ON."[61] In any case, while Project Xanadu inspired a generation of sophisticated desktop hypertext systems,[62] and the Web became the ubiquitous—if limited—networked system we use today, it was the wiki that made a dynamic and versioned Internet hypertext system widely available.

Project Gutenberg

A second line of digital lineage originates in Project Gutenberg. Started roughly at the same time as Xanadu, Gutenberg's mission is the provision of free ebooks. Whereas Xanadu was focused on innovative ways of interfacing with information, Michael Hart, a student at the University of Illinois, started Gutenberg in 1971 to provide online access to existing print information. The story of its birth is rendered in almost mythical terms. Through friends Hart gained access to a Xerox Sigma Five mainframe computer at the university's Materials Research Lab; such a machine was extraordinarily expensive, and consequently, access to it was a valuable privilege. In fact, many of Project Gutenberg's introductory materials stress that such access was worth hundreds of thousands if not millions of dollars: "At any rate, Michael decided there was nothing he could do, in the way of 'normal computing,' that would repay the huge value of the computer time he had been given . . . so he had to create $100,000,000 worth of value in some other manner."[63]

Envisioning a time when computers would be widely accessible—indeed, this computer was one of the first twenty-three that would become the Internet[64]—Hart began typing in a copy of the United States Declaration of Independence he happened to have in his backpack. And so "Project Gutenberg was born as Michael stated that he had 'earned' the $100,000,000 because a copy of the Declaration of Independence would eventually be an electronic fixture in the computer libraries of 100,000,000 of the computer users of the future."[65] Beyond being one of the first free publicly accessible cultural resources on the Internet, Project Gutenberg is relevant to the history of the universal encyclopedic vision and Wikipedia for two additional reasons.

Initial contributions to Project Gutenberg were like Hart's inputting of the Declaration of Independence: a single contributor typing in the whole text. Whereas in one profile of the project it is claimed that Hart typed in the first one hundred books,[66] Hart recalls,"I had plenty of help, even back in those days, though it was mostly anonymous, and even _I_ did not know who typed most of the first dozen or two that I didn't do."[67] This was laborious work, and in time the majority of texts being submitted were scanned and interpreted by optical character recognition (OCR) software. Yet this is an imperfect technology because books age and typefaces can be varied. The greatest challenge to Project Gutenberg was how to apportion and coordinate the work of volunteers who might have enough time to correct a chapter's worth of work, but not a whole book. In 2000, Charles Frank launched Distributed Proofreaders, a complementary project to Gutenberg that would "allow several proofreaders to be working on the same book at the same time, each proofreading on different pages."[68] Each page of the work undergoes two proofreadings that are reconciled by a "post-processor." The importance of distributed proofreading is that it permits massive collaboration. Research on Free and Open Source Software (FOSS) development has identified this characteristic of content production as asynchronous and incremental "micro-contributions."[69] It is also cumulative, as Ted Nelson noted about hypertext: data can be "reorganized constantly without losing their previous organization" and therefore "order becomes cumulative—unlike most computer systems."[70] Indeed, Distributed Proofreaders' maxim is "a page a day"—but on average readers proof more than that. This feature of allowing many contributors to produce overlapping work in bite-sized chunks—though often becoming a consuming passion—is a powerful motif in what I call *open content communities*, those communities that openly produce software and other content, such as Wikipedia.

Project Gutenberg was also responsible for one of the first publicly available reference works on the Internet, or at least part of it: volume 1 of the 1911 *Encyclopædia Britannica* (EB11). In January 1995 Project Gutenberg published the first volume of EB11, which had passed into the public domain. However, the work then stalled. When the question of resuming was raised at the end of 2002, the resulting discussion touched on the difficulties of the work including where to get the source material, how to represent textual structure, whether to preserve illustrations, and how to deal with font difficulties. Most obviously, the project would have to accommodate the size of the text, both in the amount of material, which no one would want to type, and page size, since not all editions could fit in most scanners of the time.[71] (Unfortunately, efficiently scanning books often required "destructive scanning," or removing the pages from the binding, an unpalatable task to perform on historic editions. Google's scanning project uses cameras from multiple perspectives to create a 3-D model of a naturally opened book, and then "de-warps" the image of the page so it appears flat.[72]) Yet, the work on EB11 was resumed by Distributed Proofreaders when part 1 of volume 2 was posted with much fanfare in October 2004. Continuing with the mythic character of its origins, this event was characterized as the long-awaited return to an ancient struggle:

> On the morning of October 8, 2004, near his library window overlooking a quiet lake in upstate New York, David Widger ran a series of final checks and verifications on a partitioned element of the 11th edition of the Encyclopædia Britannica. Yes, that same EB11 which has long been known as a formidable processing challenge throughout the Project Gutenberg community. This latest approach towards its digital conversion did little to diminish that reputation. . . . This "slice" of EB11 was not simply another single project being posted to the PG shelves, but the final component in a varied and impressive collection [that marked] the completion of Distributed Proofreaders' 5,000th unique title produced for Project Gutenberg and the digital public domain.[73]

Outside of Project Gutenberg, questions of how to incorporate EB11 into Wikipedia—and even the Interpedia, a pre-Web predecessor—also proved difficult. While Gutenberg had not yet completed the task, in 2002 all twenty-nine volumes of EB11 were published at http://1911encyclopedia.org/. Some saw this as an opportunity to populate Wikipedia with high quality materials: EB11 was considered one of the best references of Western knowledge at the start of the twentieth century, even if rather dated by its end. Yet copyright, trademark, and substantive issues were to hinder any efforts to make use of this online version. The organization that published the twenty-nine volumes online claimed a copyright in the work it posted, arguing that its edition was an improvement upon a public domain work. Additionally, even if the text was now in the public domain the name *Encyclopædia Britannica* remained a trademark. For this reason, the Project Gutenberg version is referred to as the Gutenberg Encyclopedia. Yet even the terms of the Gutenberg Encyclopedia proved to be confusing to some Wikipedians who wished to cite the source of the work (*Britannica* or Gutenberg) without violating trademarks and their associated licenses. And substantively, some thought that any material from a 1911 work was of little use, even for historical subjects. While some material was imported as a starting point for subsequent editing, these difficulties and the extraordinary growth of home-grown content on Wikipedia rendered the issue relatively moot.[74]

Aside from the two obvious connections between Project Gutenberg and Wikipedia, there is a lesson here central to a theme of this chapter. A strength of Project Gutenberg was that the simple vision of sharing accessible ebooks was directly satisfied by technology available at the time: one could type existing public domain books into a networked computer using "plain vanilla ASCII." ASCII is the legacy standard for representing the Roman alphabet and Arabic numerals; it was the character set used by most early computer and network developers; it is still in use today. However it has no representation for accented characters, much less non-Roman scripts. Also, a file of ASCII characters is rather sedentary. No underlines, italics, or boldface—Project Gutenberg represents all of these as uppercase. Nor does ASCII accommodate links or other hypertextual innovations.

The term *plain vanilla ASCII*, is repeated in full, like a mantra, in Project Gutenberg materials. Michael Hart was well known for his opposition to any exclusive reliance upon more sophisticated textual representations such as PDF or HTML: documents, with few exceptions, must at least be available in "plain vanilla ASCII" which could then be complemented by other formats.[75] While frustrating to some, this insistence may have prevented the project from becoming ensnared in endless debates about formats and permitted it to achieve the success it has. As one Gutenberg participant put it:

> The heart and soul of project gutenberg is the plain-text file. over the years, it has been scorned and even attacked outright. some people say it's ugly. and it's far too low-tech for others. but somehow, it has survived and even thrived in a way that no other e-book technology ever has. in the process, i have grown to appreciate its tenacity, and grasp its inner beauty. this thread is for those having a love-affair with plain-text.[76]

However, this success had not been able to yield a complete and free online encyclopedia.

Interpedia

Unlike Project Gutenberg, the Interpedia project was conceived of as an encyclopedia, but this conceptualization was confused by a plethora of technical options. The Interpedia Frequently Asked Questions (FAQ) document introduces the project by noting a resurgence in the early 1990s of the notion of a freely accessible encyclopedia:

> According to Michael Hart the idea for a net encyclopedia has been around nearly as long as the net, at least back to 1969–71. This recent burst of activity is the result of a post to several newsgroups by Rick Gates with his idea to write a new encyclopedia, place it in the public-domain, and make it available over the Internet. Among the first responses to Rick's message was one by Gord Nickerson who suggested that this Internet Encyclopedia be fully hypertexted using a markup language such as html.[77]

In October 1993, when the project was proposed by Rick Gates on the alt.Internet.services Usenet newsgroup,[78] Internet usage was reaching a critical mass. Nontechnical members of universities and technology companies were beginning to use email and Usenet. Computer hobbyists who typically communicated via dial-up bulletin board systems were developing Internet gateways so they too could access the Internet. And most importantly, new applications, and their network protocol and document formats, were proliferating. In addition to FTP (file transfer protocol) for transferring and storing files, email correspondence, and Usenet discussion groups, three new technologies were vying to be the next prominent Internet service. WAIS (Wide Area Information Server) retrieved documents based on keyword queries.[79] Gopher permitted one to browse information using menu traversal, to dig down into a publisher's taxonomy from general to specific. And, of course, there is the Web. The conundrum of which system to use is apparent in Interpedia's FAQ:

> The gopher system is widely available but is not sufficiently easy to use to satisfy many people, and it does not support hypertext. Perhaps gopher software could be improved, but it doesn't seem appropriate yet. The WWW has many advantages over earlier approaches (e.g. gopher), but is not to everyone's liking. Many people do not like navigating around in hypertext, and insist that an encyclopedia must provide keyword and/or alphabetical access. Perhaps the WWW could be improved to support the Interpedia project, but it doesn't seem quite appropriate yet. It might be a good starting point though.[80]

It is important to remember that while it is difficult today to conceive of using the Web without a search engine, this was not an integral part of the original Web when search engines still were limited, experimental services. In addition to the confounding array of options, Doug Wilson, maintainer of the Interpedia FAQ, wrote, "the term Interpedia is ambiguous—to some it means the text, to some software, and to others what we will have when we have both." A consequence, in part, of this technical uncertainty was an ambiguity in vision. Would Interpedia be part of the Internet, or, if it references existing services, would it be something "that ends up *being* the net"? This confusion is further demonstrated in answer to the question on other parallel projects to Interpedia, including FAQ documents, FTP- and Gopher-based resource guides, collections of electronic art, and the Web itself.[81]

For about half a year Interpedia participants were relatively active on the mailing lists and Usenet group. Yet, perhaps because of these ambiguities and the explosive growth of the Web, the project never left the planning stage. Even so, it is of interest for three reasons. First, in response to the hypertextual identity crisis (i.e., are we part of the Web, or is the Web part of us?), project participants envisioned at least a core or default set of encyclopedic articles. Articles could be submitted by anyone and quality and legitimacy would be arbitrated by a collection of decentralized seal-of-approval systems. No acceptance or rejections were necessary; instead, a seal "indicates that some article is good" and would be used by both people and the software to govern the accessibility of articles.[82] Second, the EB11 also proved to be a source of controversy as a strategy for initially populating Interpedia. (Michael Hart was an Interpedia member and other members eagerly anticipated all twenty-nine online volumes of the Gutenberg Encyclopedia. As noted, the first and only volume of the 1990s was posted in January 1995). Third, the process and culture of Interpedia would be facilitated by editors, whose responsibilities were "to act in good faith in the advancement of the Interpedia."[83] This notion of contributors acting in good faith anticipated a cultural norm that I argue is central to Wikipedia's collaborative culture.

Distributed Encyclopedia

Though the actual Interpedia project fizzled, its newsgroup continued to be a forum for the occasional question or announcement for years to come. The notion of an Internet-based encyclopedia was no longer novel, and as the 1990s progressed the Web became the obvious platform for any such project. In hindsight, the formation of such a reference work seems inevitable. Yet, at the time, there was little clarity on how such a project would work. Ideas and half-starts came and went—or as Foster Stockwell, a historian of reference works, noted in explaining why he didn't concern himself much with online works in 2001, they are "here today gone tomorrow."[84] In 1997, Jorn Barger posted a message entitled "Beyond the Interpedia" to the newsgroup. He wrote, "from time to time, people ask if the Interpedia project—to get a full, free Encyclopedia on the net in some form—is still happening anywhere."[85] The "closest descendent" known to Barger was the Distributed Encyclopedia.

Beyond this newsgroup posting, there are very few references to this project on the Web today. Its project pages themselves can only be found in the Internet Archive and do not give the impression of being more than a manifesto of a very small, if not single, number of authors.[86] Still, the project's introduction clearly reflects a stabilization in how such a project was conceived. It would benefit from many contributions and it would be distributed, meaning there would be no central authority (beyond simple stylistic conventions) or repository: each article would be hosted by the author and linked via a central index.

The irony here is that while it became clear that the Web would play a fundamental role, and an enormous strength of the Web is its hypertextual and decentralized character, Wikipedia itself is not decentralized in this way. It is not a collection of articles, each written by a single author, strewn across the Web. Instead, many authors can collaborate on a single article, stored in a central database that permits easy versioning, formatting, and stylistic presentation. Furthermore, there is a vibrant common culture among Wikipedians that contributes to Wikipedia's coherence.

Nupedia

In January 2000, a few months prior to the first emails to the Nupedia list with which I opened this chapter, Larry Sanger emailed Jimmy Wales with a proposal. Wales, an Internet enthusiast since his days of playing in multi-user-dungeons (MUDs) in college,[87] had been toying with the idea of an Internet encyclopedia. When Sanger emailed him about a bloglike successor to "Sanger and Shannon's Review of Y2K News Reports"—Y2K passed without much incident and both Sanger and Shannon were looking for new (funded/sponsored) activities[88]—Wales counter-proposed his encyclopedia idea and asked Sanger if he would be interested in leading the project. Each man's career path made for a fruitful collaborative potential. Wales obtained bachelor's and master's degrees in finance and took courses in the Ph.D. programs at the University of Alabama and Indiana University, but never wrote a dissertation; he instead turned to the marketplace as a futures/options trader. During the explosive growth of the Internet, Wales also began investing in, and founded his own, Internet business. Sanger was a doctoral candidate in philosophy finishing his dissertation on "Epistemic Circularity."[89] (This topic was to influence Sanger's approach to addressing issues of bias and neutrality in both the Nupedia and Wikipedia.) Both men were well educated, comfortable with technology, familiar with the norms of online community and discussion, and between them had the financial, philosophical, and academic resources to launch and sustain such a project.

In February 2000, Sanger moved to San Diego to start work at Bomis, Wales's Internet portal company. In the months before the March 9 public announcement, Sanger drafted many ideas and policies in discussion with

Wales and another Bomis partner, Tim Shell, about how to run Nupedia.[90] In the March 10 *PC World* article about the launch, the project was presented as ambitious and in need of contributors:

> The site's managers are seeking contributors and editors with expertise in, well, almost anything. The contributors will provide the diverse content, which will be offered free of charge to both consumers and businesses. Anyone is welcome to peruse Nupedia, and any other Web site may post Nupedia's content on its own. They need only to credit Nupedia as the source.[91]

The article also notes that Nupedia was inspired by other open source projects like Linux and the Open Directory Project; the goal was to be open to all expert contribution and free of charge to all users, and Sanger's quoted aspiration was for Nupedia to become "the world's largest Encyclopedia." Similarly, the signature appended to the very first email sent to the Nupedia list states "Nupedia.com building the finest encyclopedia in the history of humankind."[92]

Unlike the Interpedia—and certainly the Distributed Encyclopedia—Nupedia shows the benefit of the resources of Wales (Bomis) and efforts of Sanger. Wales wrote to the Nupedia list: "The company behind Nupedia, Bomis, Inc., has a great deal of experience designing and promoting high-traffic websites. We intend to put that experience (and the profit from that!) behind the Nupedia project to insure that it is a success."[93] In the course of its first year in 2000, Sanger was the picture of frenzied cheerleading activity. In March, Sanger reported the project had 602 members and of the 140 who had filled out membership forms "about 25–40% of these (or 35–56) are Ph.D.'s or otherwise clearly bona fide experts."[94] By the summer the first article (on atonality) was formally published and the Advisory Board was in place. By November version 3.31 of the Nupedia.com "Editorial Policy Guidelines" was published.[95] Software was frequently updated throughout the year. And, throughout, Sanger was always trying to recruit new members, including the offering of T-shirts and coffee cups, and an end-of-year membership drive with cash prizes. By January 2001 there were approximately two thousand people on the Nupedia email list.[96]

Despite these efforts and progress, Nupedia was struggling. The recruitment efforts are evidence of the difficulty in procuring commitments from volunteers for the significant work entailed in writing an article and seeing it through the complex Nupedia editorial process.[97] The universal vision, this time in the form of a "dream" of a low-cost encyclopedia available to "schoolhouses across the world" seemed reasonable, certainly compared to earlier hopes for world peace. The technology, too, seemed capable of inexpensively supplying information throughout the world, and even facilitating the work of distant contributors. Yet something more was needed and it would only be found by (seeming) accident. But before I turn to the wiki, there's one more stop.

GNUPedia/GNE

In January 2001, the same month in which the Nupedia mailing list had reached approximately two thousand subscribers, a controversy erupted around a Slashdot posting entitled "Will The Real Nupedia Please Stand Up?"[98] Richard Stallman, father of the Free Software movement, which itself was an inspiration for Nupedia, announced a competing project led by Hector Arena. Under the aegis of Stallman's GNU organization the GNUPedia would implement a proposal Stallman had drafted in 1999 for a "free universal encyclopedia and learning resource." (GNU is a recursive acronym for "GNU's Not Unix" and it set out to replace the proprietary Unix operating system with a similar but free system.) Stallman's proposal for a "free universal encyclopedia" had been presented in various venues in 1999 (e.g., the SIGCSE conference in March and the MacArthur Fellows Reunion in October[99]), but only came to be known publicly when it was made available on the Web as part of the controversial GNUPedia project announcement in 2001. Stallman outlined a vision of single-author articles distributed throughout the Web but indexed by the central project[100]—much like the Distributed Encyclopedia. This vision purposely eschewed any type of central authority besides a commitment to freedom, meaning any article to be linked must satisfy the criteria of permitting universal access, mirroring, modification, translation, and quotation with attribution. Given the lack of central control, these criteria would be enforced by compliant articles or indexes refusing to link to any encumbered article. Additionally, Stallman encouraged contributions from educators (whose disciplines he thought were becoming increasingly commercialized), and envisioned peer review and endorsements—similar to Interpedia seals of approval. (On the quality-ranking front, in May 2008 the German Wikipedia began using the "Flagged Revisions" feature to mark acceptable/stable versions of a page; other language editions may follow.)[101] In Stallman's proposal, the Web-like assumption of decentralization was again present. And "freedom" was ensured by the same reciprocity required by copyright licensees that govern free software: nonfree is kept separate from the world of the free. Most importantly, the proposal recognized important challenges previous projects failed to meet: contributors should appreciate that "small steps will do the job" when one "takes the long view."[102]

Even so, this humble and ambitious sentiment of the tortoise getting there in the end wasn't enough; an actual system was never realized. Because the name and the announcement were not meant to intentionally interfere with Nupedia, GNUPedia refocused as a "library of options" or "knowledgebase" and changed its name to GNE, a recursive acronym, like GNU, standing for "GNE is Not an Encyclopedia." Stallman wrote to me that this incident was a simple case of confusion as he was in discussion with multiple people about encyclopedic projects without remembering that they were distinct, but he wanted to ensure any and all such projects would respect freedom.[103] Yet while GNE project participants wrestled with their new purpose, at the same time expressing concern about the centralization and complexity of the Nupedia process, Wikipedia quickly overtook both.

The Web and Wikis

To understand the success of Wikipedia as the most credible realization of the universal encyclopedic vision, one must also understand a failing of the Web as we know it, but not as it was first conceived. In his memoir, World Wide Web inventor Tim Berners-Lee writes that his motivation was to design the Web as "a universal medium for sharing information."[104] While hypertext pioneer Ted Nelson considered the Web a hobbled upstart relative to Project Xanadu, the Web as it works today falls short of even Berners-Lee's original vision, which he now refers to in its richer potential as the "Semantic Web."[105]

In any case, despite the Web's early limitations, or perhaps because of them, in January 1993 there were nearly fifty different web browsers.[106] These were inspired by Berners-Lee's original Web client and roughly implemented the specifications for HTTP (network transport), HTML (content markup), and URL (resource locators/identifiers). However, one client was to stand out among others: Mosaic, which led to Netscape. Unfortunately, some Mosaic developers were seemingly intent on overshadowing the World Wide Web and failed to implement the critical feature of editing a Web page. Berners-Lee writes, "Marc and Eric [Mosaic developers] explained that they had looked at that option and concluded that it was just impossible. It can't be done. This was news to me, since I had already done it with the World Wide Web [client] on the NeXT—though admittedly for a simpler version of HTML."[107]

Consequently, for many people the Web became a browsing-only medium unless they were savvy enough to know how to manually publish Web pages, or fortunate enough to use a fully featured Web client such as Arena or AOLPress. Until, that is, the WikiWikiWeb. As already noted, "wiki wiki" means "super fast" in the Hawaiian language, and Ward Cunningham chose this name for his wiki project in 1995 to indicate the ease with which one could edit Web pages. Wiki makes this possible by placing a simple editor within a Web page form, with formatting and linking functions carried out by the wiki server.

At the beginning of January 2001 frustration increased over Nupedia productivity. The need to publish more articles, as well as a greater popular interest in contributing, was not well matched by the expert-dependent multistep editorial process. Hence, the stage was set for the introduction of a wiki. On January 2, at a San Diego taco stand, Sanger had dinner with Ben Kovitz, a friend from Internet philosophy lists, during which Kovitz introduced the idea of wikis to Sanger.[108] (The background of wiki is further discussed in the next chapter.) The wiki could be a possible remedy to Nupedia's problems, permitting wider contribution and collaboration on articles that would then be fed to Nupedia's editorial review. Within a day, Sanger proposed the idea to Wales and Nupedia's wiki was announced on January 10 in a message entitled "Let's make a Wiki":

> No, this is not an indecent proposal. It's an idea to add a little feature to Nupedia. Jimmy Wales thinks that many people might find the idea objectionable, but I think not. . . .
>
> As to Nupedia's use of a wiki, this is the ULTIMATE "open" and simple format for developing content. We have occasionally bandied about ideas for simpler, more open projects to either replace or supplement Nupedia. It seems to me wikis can be implemented practically instantly, need very little maintenance, and in general are very low-risk. They're also a potentially great source for content. So there's little downside, as far as I can see. . . . If a wiki article got to a high level it could be put into the regular Nupedia editorial process. . . . On the front page of the Nupedia wiki we'd make it ABSOLUTELY clear that this is experimental, that Nupedia editors don't have control of what goes on here, and that the quality of articles, discussion, etc.,

should not be taken as a reflection of the quality of articles, review, etc. on the main part of the Nupedia website.[109]

However, Nupedia contributors resisted Nupedia being associated with a Web site in the wiki format. Therefore, the new project was given the name "Wikipedia" and launched on its own address, Wikipedia.com, on January 15, 2001.[110]

Wikipedia

Since its start, Wikipedia's growth has been extraordinary. Within six months Sanger announced that "the Wikipedia is now useful."[111] In September he proclaimed on Usenet that the "Interpedia is dead—long-live the Wikipedia": "Interpedia's noble dream of creating a free, open encyclopedia lives on—not quite in the form imagined, but in a 'very' open and free form with which many early participants would probably approve."[112] Wikipedia proved to be so successful that when the server hosting Nupedia crashed in September 2003 (with little more than twenty-four complete articles and seventy-four more in progress) it was never restored.[113] As already mentioned, there are now scores of active language encyclopedias, millions of articles, and a handful of other Wikimedia projects. There are, of course, thousands of other wikis, many quite specialized and a few continuing forward with the universal vision. For example, the home page of the (seemingly dormant) Collective Problem Solving Wiki strikes me as true to the aspirations of H. G. Wells: "Our world has complex and urgent problems that need to be addressed. We believe there are innovative ways of solving them together online."[114]

And while Wikipedia is a remarkable realization of a century-old vision, the end of this story is not as happy as it might otherwise be—nor is it really the end, just where I finish this part of the tale. In the first year of Wikipedia's life, its radical openness and explosive growth were never reconciled with Nupedia's goal of an authoritative expert-driven reference work. Once it was clear that a wiki could be useful, Sanger tried to introduce the idea again for Nupedia:

But by the summer of 2001, I was able to propose, get accepted (with very lukewarm support), and install something we called the Nupedia Chalkboard, a wiki which was to be closely managed by Nupedia's staff. It was to be both a simpler way to develop encyclopedia articles for Nupedia, and a way to import articles from Wikipedia. No doubt due to lingering disdain for the wiki idea—which at the time was still very much unproven—the Chalkboard went largely unused. The general public simply used Wikipedia if they wanted to write articles in a wiki format, while perhaps most Nupedia editors and peer reviewers were not persuaded that the Chalkboard was necessary or useful.[115]

Stretched between continuing frustration with Nupedia's progress, problems with unruly Wikipedians, and a widening gap between the two, Sanger failed to save Nupedia and alienated some Wikipedians who saw his actions as increasingly autocratic.[116] Additionally, with the burst of the Internet bubble, Sanger, among many others in the industry, was laid off from Bomis and resigned from his Wikipedia role shortly thereafter. Sanger's subsequent commentary from the sidelines, particularly his continued criticism of Wikipedia not respecting the authority of experts, prompted additional negativity toward him. In April 2005, Sanger published his memoirs of Nupedia and Wikipedia, which sparked a controversy over whether Sanger even deserved credit as a cofounder of Wikipedia.[117] In March 2007 Sanger launched a new encyclopedic project, Citizendium, with the intention of improving "on the Wikipedia model by providing 'reliable' and high-quality content; . . . by requiring all contributors to use their real names, by strictly moderating the project for unprofessional behaviors, and by providing what it calls 'gentle expert oversight' of everyday contributors."[118] In early 2009, in a sad irony for a project based on good faith, the question of credit and cofounding erupted again, with Sanger and Wales becoming even more embittered and accusing each other of dishonesty.[119] Sanger's exit from Wikipedia will be further touched on when I consider leadership in such communities, as will the larger social debate about "experts versus amateurs."

Conclusion: Predicting the Future, Reading the Past

A history professor of mine once wisely noted "historians stink at predicting the future." Predictions about technology, regardless of who makes them, seem especially problematic.[120] Even those who help "make" the future are no better at prediction. In this chapter I considered those looking back, those looking forward, and those struggling in their present to implement a universal encyclopedic vision. For a long time, no one got it quite right. But people, being people, try, and try again. And that story is revealing in at least two ways.

First, even unfulfilled visions, failed projects, and erroneous predictions tell us something about those people and their time. The history recounted in this chapter speaks to the alluring and enduring notion of an ambitious project of human knowledge production and dissemination: a universal encyclopedia. This vision persisted throughout the twentieth century even though each instance was prompted by different technologies and entailed differing levels of accessibility in production: Otlet's documentalists, Wells's technocrats, Nupedia's scholars, and Wikipedia's "anyone."

Second, a question throughout this chapter is why did it take so long for the vision to be realized? A possible answer can be detected in the overlapping spheres of vision, pragmatics, and happenstance; interesting things happen when those stars align. Perhaps the best example of this can be seen in the expectation (i.e., of the Distributed Encyclopedia and GNUPedia) that once it was clear the Web would be a platform for such an encyclopedia, it would also be decentralized. But, Wikipedia is centralized, in part, because wikis made editing the Web possible again for many people—and the loss of editing capabilities from Berners-Lee's original vision was seemingly another chance event. Wikis have other features that make them useful for an encyclopedia (e.g., versioning and simple inter-wiki linking)—though, seemingly, Wales himself thought such a notion would not be received well and Ward Cunningham predicted that the result would be more a wiki than an encyclopedia[121]

In any case, the projects discussed in this chapter are attempts at realizing a universal vision, encompassing the goodwill of collaborators and reaching toward global accord. While it is a mistake to argue all reference works are necessarily progressive, as I warn in chapter 7, even *Britannica*—often thought to be the conservative opponent of the *Encyclopédie* in the 1800s and Wikipedia today—shared this sentiment of global accord in a preface to a 1956 edition of its world atlas: "To the men, women, and children of the world who, by increasing their knowledge of the earth and its people, seek to understand each other's problems and through this understanding strive for a community of nations living in peace, the *Encyclopædia Britannica* dedicates this volume."[122] And while few would argue that Wikipedia will necessarily further world peace, in the next chapter I argue "good faith" culture is necessary to its production and an occasional consequence of participation.

A Timeline of Events

1895 Otlet's Permanent Encyclopedia: liberating ideas from the binding of books.

1936 Wells's World Brain: a vision of a worldwide encyclopedia using microfilm.

1945 Bush's memex: a vision of a hypertextual knowledge space and new forms of encyclopedias.

1965 Nelson's Xanadu: a vision of hypertext.

1971 Hart's Project Gutenberg: a vision of providing ebooks through achievable means ("plain vanilla ASCII").

1980s Academic American Encyclopedia is made available in an online experiment; multimedia CD-ROMs soon follow.

1991 Berners-Lee's World Wide Web: a vision of highly accessible read/write.

1993 Interpedia: an ambiguous vision lost among too many infrastructural options.

1995 Cunningham's WikiWikiWeb: making the Web easy to edit collaboratively.

1999 Distributed Encyclopedia: many people should contribute independent essays that could be centrally indexed.

1999 Stallman's "The Free Universal Encyclopedia and Learning Resource."

2000 Distributed Proofreaders: distributing the task of proofreading among many.

2000 (March 9) Nupedia launched: a FOSS-inspired expert-driven free encyclopedia.

2001 (January 10) "Let's make a wiki."

2001 (January 15) www.wikipedia.com launched.

2001 (January 16) GNE project announced.

2001 (September) "Interpedia is dead—long live the Wikipedia."

3 Good Faith Collaboration

All rules and guidelines add up to this; Respect!
—Phoenix 15's Law

There are two complementary postures at the heart of Wikipedia collaboration: the stances of "Neutral Point of View" (NPOV) and good faith. Whereas other communities may have a culture of good faith (i.e., assume good faith on the part of others, and act with patience, civility, and humor), few are concerned with producing an encyclopedia. The dovetailing of an open perspective on knowledge claims (epistemic) and other contributors (intersubjective) makes for extraordinary collaborative potential, and harkens back to the universal vision of increased access to information *and* social accord. Furthermore, perhaps an understanding of neutrality and good faith can serve as a rejoinder to a favorite quip about Wikipedia, also known as its Zeroeth Law: that while it may very well work in practice, it can never work in theory.[1]

Introduction

Before engaging with the Wikipedia's collaborative culture, it is worthwhile to frame such an undertaking. (Again, my focus is on the English-language Wikipedia; comparative work between Wikipedias in other languages does show differences in conception of power, collectivism, and anonymity.[2]) I begin this introduction at the most abstract level by briefly explaining what I mean by "collaborative culture." I also note that there is often a disconnect between written policy and actual practice within organizations; in offering a bit of history about how wikis came to be, I argue wikis help close the gap between policy and practice. I then explore the background, theory, and practice of neutrality and good faith by way of a conflict about the English Wikipedia's "Evolution" article.

A Caveat about Collaborative Culture

Heretofore I have used the term *collaborative culture* in a commonsensical manner, but if pressed for further explanations on what collaboration or culture mean one can find many and varied answers. Indeed, authors have commented on the variety of approaches to "culture" across disciplines, including anthropology, communications, and history.[3] Within the realm of organizational studies Edgar Schein posits eleven different categories of how culture is commonly conceived. In this project, I speak of culture as the "way of life of a people,"[4] the value-laden system of "meaning making" through which a community understands and acts, including its own maintenance and reproduction. Schein writes that "culture acts as a set of basic assumptions that defines for us what to pay attention to, what things mean, how to react emotionally to what is going on, and what actions to take in various kinds of situations."[5]

Similarly, *collaboration* can be an equally provocative term prompting debate, for example, about the difference between coordination or cooperation and collaboration.[6] Additionally, collaboration stands among other related concepts such as dispute resolution, conflict management, and interdependent decision making. Each of these notions, and their literatures, are useful but, alone, insufficient. For example, the notion of "dispute resolution" is surprisingly optimistic, as if agreement and harmony are the natural state from which disputes sometimes errantly arise and must be swiftly corrected. Yet to characterize social relations as inherently conflicted—as when Wikipedia is humorously characterized as an "argument engine"[7]—is also mistaken. Nor is conflict necessarily a bad thing: legal scholar Cass Sunstein convincingly argues that dissent is a critical and generative contribution to society.[8] For this reason, recent textbooks on the topic prefer conflict "management" to "resolution" and recognize that consensus and dissensus each have an important, and unavoidable, role in community. In this way Wikipedia is like the Free and Open Source Software (FOSS) communities as characterized by Steven Weber:

> The open source software process is not a chaotic free-for-all in which everyone has equal power and influence. And is certainly not an idyllic community of like-minded friends in which consensus reigns and agreement is easy. In fact, conflict is not unusual in this community; it's endemic and inherent to the open source process.[9]

Recognizing this, we may instead wish to refer to "interdependent decision making,"[10] which appropriately shifts the connotation away from "conflict-is-bad." However, much more is involved in Wikipedia production than decision making. Consequently, I use the term *collaboration* in Michael Schrage's sense, which arose from his study of collaborative technologies: "collaboration is the process of shared creation: two or more individuals with complementary skills interacting to create a shared understanding that none had previously possessed or could have come to on their own. Collaboration creates a shared meaning about a process, a product, or an event."[11]

Therefore, my use of the term *collaborative culture* refers to a set of assumptions, values, meanings, and actions pertaining to working together within a community. And, in many ways my use is like that of media scholar Henry Jenkins's notion of "participatory culture" in which consumer-only fans of commercial genres (e.g., sci-fi) are now creators within their own "fandom" communities. Jenkins defines participatory culture as one in which there are low barriers of engagement, support for creation and sharing, and some form of mentorship or socialization, and members believe that their contributions matter and they "feel some degree of social connection with one another."[12] By these criteria, Wikipedia would qualify.

Wiki, Practice, and Policy

Douglas Engelbart, a father of the modern computer interface, wrote in his essay "Augmenting Human Intellect" that computers would permit researchers themselves to benefit from the product of their work through a regenerative "feeding back of positive research results to improve the means by which the researchers themselves can pursue their work."[13] More than forty years later anthropologist Christopher Kelty observed this phenomenon among technical communities using the Internet. Likely unaware of Engelbart's prediction, Kelty chose to call such communities a "recursive public": a form of "social imaginary" through which geeks collectively conceive their "social existence" and are capable of changing the very means of discourse (i.e., communication protocols).[14] I can think of no better example of this notion of "regenerative" or "recursive" feedback than Wikipedia.

To understand why, consider another complementary notion, Etienne Wenger's "community of practice," developed with Jean Lave. In this theory people are understood to pursue a shared enterprise over time yielding a common identity and understanding of their environment; they accumulate a rich repertoire of cultural norms and actions. In addition to actual participation/practice, Wenger's theory provides for reification: "the process of giving form to our experience by producing objects that congeal that experience into 'thingness.'"[15] Whereas others have cast wikis as communities of practice,[16] I find one of the most interesting facets of the theory to be the relationship between practice (e.g., creating an encyclopedia) and its "reification" (e.g., documenting the community's practice). Wenger argues that practice and reification are not opposites, but coexist in a "duality of meaning" of interaction and interplay.[17] However, in many traditional projects and organizations the documentation of organizational culture and process (i.e., reification) is often dramatically out of step with actual practice. But the wiki can change this.

Wikis were born of an advocacy for a change in software development with respect to how application requirements were perceived (i.e., as patterns) and satisfied (i.e., agilely). In the 1990s a new way of addressing software requirements was becoming popular: the "design pattern." Rather than confronting every new task as a new problem to be solved, it was believed that experience could be distilled into a shareable set of design patterns. (A pattern is a higher-level abstraction than that of the computer algorithm, which is a common way of addressing a particular computational task, like sorting a list.) For example, a software engineer might be confronted with a task in which a service acts on behalf of another. This might be an instance of the "proxy pattern" that might already be well understood. Ward Cunningham, an advocate of design patterns, attended a conference on pattern languages where he agreed to collect and post user-submitted patterns if contributors sent him a structured text file that he could then automatically process and post online. This was surprisingly difficult for many: "And I was amazed at how people who sent me files couldn't follow even the simple rules. I was three pattern documents into this thing, and getting pretty tired of it already. So I made a form for submitting the documents."[18] This user-editable repository, started in 1995, would come to be known as the Portland Pattern Repository and the first wiki.[19]

Furthermore, requirements, often perceived as patterns, would be satisfied differently too. Unlike earlier software development in which all requirements for a project were carefully collected and completely specified, and only then implemented, "agile software development" advocates argued these steps should be collapsed and iterated in small increments. Instead of a large collection of requirements going out of date, requirements are often specified as a set of user scenarios and related test cases that can be objectively satisfied and tested for regressions—to prevent fixes and new features from creating new bugs. The authors of the "Manifesto for Agile Software Development," including Ward Cunningham, wrote that they valued: "Individuals and interactions over processes and tools. Working software over comprehensive documentation. Customer collaboration over contract negotiation. Responding to change over following a plan."[20]

A benefit of this approach is that at each step there is always some working code satisfying the requirements encountered so far, and the software is easily extended and adapted as requirements change, as they are bound to do. However, there was still a need for quickly, flexibly, and collaboratively discussing software, design patterns, and the principles of this new paradigm. The wiki, evolving from Cunningham's user-editable pattern repository, satisfied these needs well, and over time became a useful documentation tool for many others, including those attempting to write an encyclopedia. In fact, the ability to easily document one's world satisfies a deep need in some Wikipedians, again placating the fear that doom might be averted if we learn from our mistakes:

> [W]e need to document best practices, both for new people and for old people, so that we know what we are doing. If we do not document, we cannot learn from our history, and are doomed to repeat it.
>
> The fact that one must document, document, document is ingrained in my psyche (I'm trained as a scientist, and work as a programmer). It is almost impossible for me to understand a world where documentation does not exist. . . . —Kim Bruning (talk) 20:26, 15 May 2008 (UTC)[21]

As we learned, early documentalists made great use of the index card; Ward Cunningham has also spoken about how useful index cards were to him. In his Wikimania 2005 keynote speech "Wikis Then and Now," Cunningham noted that a piece of software he used when first thinking about software patterns and human collaboration was HyperCard.[22] This Apple application was a popular hypertext system before the Web and relied on the metaphor of stacked index cards. However, Cunningham wanted a messier system in which one could talk about and refer to something that did not formally exist yet, hence the famous "red link" on wikis that points to a page not yet filled with content.[23] Furthermore, he began to use real index cards when meeting with collaborators. Index cards proved a useful way for people to talk about their processes and requirements: one could spread cards on the table, write on them, and pass them around with others—serving as what the knowledge management literature refers to as boundary (spanning) objects.[24] People would ask him: "help us find our objects" and handling the cards prompted information sharing between participants regardless of their status within the organization. Furthermore, like a red wiki link, people would often point to a blank area on the table where the nonexistent (not yet defined) card would eventually go: "They had need for a name for something they didn't know how to say."[25] It's striking that the index card, a source of inspiration from the beginning of the twentieth century, would inspire hypertext, which in turn would inspire use of the physical card, and then a new type of hypertext.

While it is increasingly difficult to find in Wikipedia articles, the red link does still exist, inviting others to fill in a bald spot of encyclopedic coverage. There is also the "stub," one step up from the red link, an article with little more than a few sentences or paragraphs. Author and commentator Nicholson Baker considers the stub to be one of the most charming features of Wikipedia collaboration, likening it to an "unusually humble" ask for help.[26] And, not surprisingly, wiki-driven editing pervades Wikipedia. That is, in addition to the encyclopedia articles, collaboratively edited using wiki, there are discussion pages about articles; pages in the Wikipedia namespace (or section) of the encyclopedia for its policy and guidelines, the Meta wiki's policy pages for all Wikimedia projects,[27] and pages for discussing changes to the underlying wiki software. (Pages in the Wikipedia namespace are frequently referred to via shortcuts, for example "WP:NPOV" refers to the NPOV policy in the Wikipedia namespace.) Each of these is wiki too. There are even third-party wikis, such as Meatball, "a common space for wiki developers and proprietors from all over the Internet to collaborate."[28] The wiki fulfills Engelbart's prediction of regenerative feedback, tightens the recursive turn of Kelty's public, and converges with Wenger's duality of meaning. Jean le Rond d'Alembert's 1751 observation about the *Encyclopédie* still appears to be true, that "since there is some incontestable advantages in being able to convey and receive ideas easily in mutual intercourse, it is not surprising that men have sought more and more to augment that facility."[29]

Scholars have posited a number of ways in which wikis facilitate this collaborative augmentation. Networking technology and its related collaborative techniques can enable openness and accessibility (e.g., discussion lists, distributed software development, and wikis), furthering accountability and the socialization of newcomers.[30] Also, people can communicate asynchronously and contribute incrementally.[31] With wikis the timing and granularity of a contribution can be as marginal as fixing a typo on a page that hasn't been touched in months. Wikis permit changes to be reverted so contributors can be bold in action and need not be brittle in response to the actions of others.[32] "Collective creation" and coordination is facilitated by persistent documentation and use of discussion pages and templates.[33] Automated tools can further aid users, and the collaboration these mechanisms facilitate is likened to "distributed cognition." For example "bots," autonomous programs, can watch edits in real-time and revert them immediately (e.g., if an edit contains profanity) or list them as suspicious. Such information can then be followed by user applications that prioritize suspicious edits based on their own heuristics, such as contributor anonymity or previous warnings, and enable single-click reversion, user warning, and administrative notice.[34] Even the ability to temporarily lock a page can be seen as a productive feature that permits the dampening of flamewars and the enforcement of cool-down periods.[35] Difficult issues in articles can be broken down: contentious material can be isolated and addressed elsewhere without impeding the progress of everything else; indeed, modularization in general is a powerful aid in interaction and content development.[36] Additionally, wikis are wonderful repositories of a community's practice and discourse. As Bo Leuf and Ward Cunningham write in their 2001 book *The Wiki Way*, "In any Wiki, you discover a sense of growing community that expresses itself through its archived writing."[37]

Wikipedia Policy, Guidelines, and the Five Pillars

In principle, there are three levels of authority associated with Wikipedia norms: *essays*, nonauthoritative pages that may contain useful insights; *guidelines*, actionable norms approved by general consensus; and *policy*, much the same but "more official and less likely to have exceptions."[38] The line of distinction between guidelines and policy is rarely bright, as evidenced in discussions about the deprecation of "Assume Good Faith" (AGF) from a policy to a guideline.[39] (A simple summary of this discussion is that AGF was rarely actionable since it involved assumptions about others' motives while "Civility" and other corollaries remain "policy" because they can be tested and enforced against more objective features of behavior.)

Wikipedia's many norms are also commonly grouped together. For example, the "Policies and Guidelines" page stresses these precepts: Wikipedia works by building consensus; Wikipedia is an encyclopedia; respect other contributors; don't infringe copyrights; avoid bias; and add only information based on reliable sources.[40] The "policy trifecta" states the three central principles of Wikipedia collaboration are as a collaborator on an encyclopedia, use a neutral point of view; as a member of a community, "don't be a dick"; and as a user of a fast and flexible wiki, "ignore all rules."[41] I find the "five pillars" to be the most complete and sensitive summary of Wikipedia collaborative norms:

> Wikipedia is an encyclopedia incorporating elements of general and specialized encyclopedias, almanacs, and gazetteers. All articles must strive for verifiable accuracy: unreferenced material may be removed, so please provide references. Wikipedia is not the place to insert personal opinions, experiences, or arguments. . .
>
> Wikipedia has a neutral point of view, which means we strive for articles that advocate no single point of view. Sometimes this requires representing multiple points of view, presenting each point of view accurately, providing context for any given point of view, and presenting no one point of view as "the truth" or "the best view." It means citing verifiable, authoritative sources whenever possible, especially on controversial topics. When a conflict arises regarding neutrality, declare a cool-down period and tag the article as disputed, hammer out details on the talk page, and follow dispute resolution.
>
> Wikipedia is free content that anyone may edit. . .
>
> Wikipedia has a code of conduct: Respect your fellow Wikipedians even when you may not agree with them. Be civil. Avoid conflicts of interest, personal attacks, and sweeping generalizations. Find consensus, avoid edit wars, follow the three-revert rule, and remember that there are 3,002,347 articles on the English Wikipedia to work on and discuss. Act in good faith, never disrupt Wikipedia to illustrate a point, and assume good faith on the part of others. Be open and welcoming.
>
> Wikipedia does not have firm rules besides the five general principles presented here. Be bold in editing, moving, and modifying articles. Although it should be the aim, perfection is not required. Do not worry about making mistakes. In most cases, all prior versions of articles are kept, so there is no way that you can accidentally damage Wikipedia or irretrievably destroy content.[42]

The first and third pillars of Wikipedia as an encyclopedia and as something "anyone can edit" will be explored in subsequent chapters. Throughout the rest of this chapter I explore the second and fourth pillars: the norms of neutrality and Wikipedia's good faith "code of conduct."[43]

Neutral Point Of View and Good Faith: An Example

One of the many contentious articles I follow on Wikipedia is that on evolution. Frequently those with criticisms of evolution, predominately religious literalists, attempt to include these criticisms in the "Evolution" article. Yet, Wikipedia articles are not forums for debate, nor are their discussion pages: "Please remember that this page is only for discussing Wikipedia's encyclopedia article about evolution. If you are interested in discussing or debating evolution itself, you may want to visit talk.origins or Wikireason."[44]

The stance of neutrality implies that contributors should abandon efforts to convince others of what is right or true, and instead focus on a neutral presentation of what is commonly understood about that topic. Consequently, much like a creationist might view the "Evolution" article, I appreciate the "Creationism" article's thorough and dispassionate treatment of the relevant history and arguments, even though I might disagree with them. Once understood and practiced, the neutrality stance permits collaboration between those who might otherwise fall into rancorous discord. Therefore, as Jimmy Wales has noted, NPOV should be understood "as a social concept of co-operation." In response to a question about objectivity and truth in Wikipedia—and the influence of Ayn Rand's Objectivist philosophy on his views—he said "The whole concept of neutral point of view, as I originally envisioned it, was this idea of a social concept, for helping people get along: to avoid or sidestep a lot of philosophical debates. Someone who believes that truth is socially constructed, and somebody who believes that truth is a correspondence to the facts in reality, they can still work together."[45]

Even so, there is still a margin for disagreement about the proportionality of even "neutrally" presented views. How much of the "Evolution" article should be dedicated to creationist objections? "Verifiability" has an important role to play here, as recognized by Gizza's First Law: "Those who believe that WP:NPOV refers to equal respect towards all verifiable perspectives are Wikipedians. Those who think that NPOV means equal coverage of all verifiable perspectives are trolls."[46] Obviously, those who cannot appreciate the relative weight of well-supported claims (i.e., the consensus of peer-reviewed research supporting evolution) will have a difficult time at Wikipedia. However, I would not actually consider such contributors as "trolls." (While the term has taken on a general pejorative function, *trolls* properly signify those who post controversial or irrelevant messages with the intention of disrupting an online community.[47])

Here, the technical feature of hypertext links can provide a calming effect. A complete treatment of evolutionary mechanisms and its history as a concept need only mention there are related "social and related controversies," which may merit their own articles. However, one should be careful in articles about controversy to avoid "content" or "POV" forking in which two articles with opposing points of view arise in place of a single NPOV article.[48] Again, in taking a neutral stance one's task is to describe the controversy rather than to partake in it.

Just as one can find contentious articles, one can also find apologies. If the stance of neutrality implies a willingness to put aside one's own "point of view," an apology is a potentially rich example of good faith. Consider the following exchange from the "Evolution" talk page. Salva31, an admirer of the conservative American columnist Patrick Buchanan,[49] became increasingly frustrated with the "Evolution" article. After Salva31's efforts to change the article were rejected, he tried to remind the scientifically minded contributors opposing him that "Wikipedia is not a battleground" and the removal of his text was not in "a spirit of cooperation." In the conversation that followed, fellow Wikipedian Branaby dawson replied:

> I'm sorry Salva but I do not think that your comments to this talk page really qualify either as in "a spirit of cooperation." I think that you have been guilty of many of those things you are accusing others of. You have broken the above rules in several ways: You've insulted people by the tone you've used in discussion. You've tried to intimidate those who don't agree with you by the shear volume of your text (on the talk page). You've not been civil or calm with your edits.
>
> As such although I have criticised others for deleting much of your text in which you do these things I would support them in moving all such material to a subpage in [the] future. Barnaby dawson 09:00, 13 Apr 2005 (UTC)[50]

While dawson's "I'm sorry Salva but I do not think. . ." isn't a genuine apology, but rather is a form of the infamous "sorry. . . but," it is nonetheless indicative of a type of discursive openness: "sorry" softens the statement, using a name promotes a sense of connection, and "I do not think" connotes a sense of fallibility. This was followed by an attempted de-escalation:

> Let's not do that. As long as Salva 31 keeps it short and simple and on topic, there shouldn't be a problem in future, right? Kim Bruning 10:30, 13 Apr 2005 (UTC)

Another participant, a graduate student in biology,[51] soon conceded to some incivility:

> Also, to be fair to Salva, I was pretty uncivil to him, I think. Graft 12:02, 13 Apr 2005 (UTC)

And within this conversation a genuine apology did manifest:

> Thank you, Graft. This is obviously a debate that is sensitive on both sides. Likewise, I owe you an apology for the contributions I made in escalating the argument. Salva31 09:37, 13 Apr 2005 (UTC)

Like many articles and discussion pages on Wikipedia, the "Evolution" article has plenty of disagreements, arguments, and even downright hostile behavior. However, NPOV policy asks editors to change their (epistemic) perspective with respect to the claims they make about the world. Similarly, the broad notion of good faith, including civility and a willingness to apologize, asks editors to extend their (intersubjective) perspective toward other contributors as well-meaning but possibly mistaken human beings.

The Epistemic Stance of Neutral Point of View

In chapters 1 and 3 I introduce the NPOV policy by way of example because it can be a confusing term. Misunderstandings about it arise in part because, as the Wikipedia article itself admits, "the terms 'unbiased' and 'neutral point of view' are used in a precise way that is different from the common understanding." People are acknowledged to be subjective beings (i.e., "inherently biased"), but when used in the Wikipedia context articles are considered to be without bias when they "describe the debate fairly rather than advocating any side of the debate." A more recent version of the page suggests one way to think about it is to "assert facts, including facts about opinions—but do not assert the opinions themselves."[52]

This notion of neutrality is also difficult because it seems impossible to explain without recourse to an equally problematic constellation of concepts. If neutral means unbiased, and unbiased means fair, might fair mean impartial, or something else? Another source of confusion is the subject of the alleged neutrality: the platform, processes and policies, people, practices, or the resulting articles? Can bias in one contaminate the neutrality of another? Additionally, the use of the prefixes *un* and *non* with words such as *bias*, *fair*, and *neutral* is indicative of one more problem. Although we might find a clear definition of what *bias* is, for example, that definition might not be as useful when we wish to understand what it means to be *unbiased*. Take, for example, the acronym POV, which has acquired a derogatory connotation as the seeming opposite of NPOV. Yet, when the acronym is expanded, to accuse someone of having a point of view seems rather ridiculous, even to those who advocate the NPOV policy.[53]

In order to bring some clarity to this, one might look to other uses of the notion of neutrality, including in gameplay, technical systems and standards, content regulation, and international conflict. From this, one can discern an understanding of neutrality as a *sensitivity* to the ways in which technical and social systems might be unfairly discriminatory; an *impartiality* and *plurality* between possible participants or positions; an ethos of *sportsmanship* and an *adherence* to known rules; and a submission to some *authority for arbitration*, as well as an expectation of *accountability*.[54] This understanding does seem to fit the personal intentions and larger aspirations of Wikipedia contribution. In the Wikipedia context the notion of neutrality is not understood so much as an end result, but rather as a stance of dispassionate open-mindedness about knowledge claims, and as a "means of dealing with conflicting views."[55]

Yet one might ask, shouldn't such a stance be the case for contributors to any encyclopedia, or even any wiki? Historically, reference works have made few claims about neutrality as a stance of collaboration, or as an end result. While other reference works have had contributions from thousands of people, they were still controlled by a few persons of a relatively homogeneous worldview. Indeed, a preoccupation of traditional references is their authoritativeness, quite different from Wikipedia's abandonment of "truth." As Nupedia's early editorial guidelines noted, "There are many respectable reference works that permit authors to take recognizable stands on controversial issues, but this is not one of them."[56] This is not to say that reference works are always regarded as being without bias: reference works have been central to many ideological battles. And pointing out the quaint biases of reference works is an amusing hobby of bibliophiles. For example, A. J. Jacobs's lighthearted diary on reading the whole of the *Britannica* notes many remnants of Victorian cultural bias (e.g., a preoccupation with explorers, botanists, and the victims and mistresses of monarchs).[57] Or, consider a Wikipedian's description of his 1898 copy of Pear's *Cyclopedia*:

It had a general encyclopedic section. I think the most wonderfully opinionated article I found in this was on Russia, which after a few breathless passages on how wonderful and civilised the place was ended with ". . . which is why Russia simply must get a port on the Mediterranean!" Extreme case, but not rare.[58]

The concept of neutrality was also absent at the birth of the wiki, which, as described, was a platform for advocating a particular type of software development. Instead, neutrality arose in the context of Wikipedia's predecessor, Nupedia, and the philosophical interests of its cofounders.

Sanger's doctoral dissertation in philosophy focused on the thorny aspects of justifying knowledge and was opaquely entitled, as they are apt to be: "Epistemic Circularity: An Essay on the Justification of Standards of Justification."[59] Wales, for his part, was not a professional philosopher, but as was not uncommon among early amateur Net philosophers, he was an Objectivist, in the Ayn Rand tradition, and moderated an email list dedicated to the topic.[60] Sanger recounts that both he and Wales were in agreement on the importance of the principle of neutrality, which was called "nonbiased" at the time:

Also, I am fairly sure that one of the first policies that Jimmy and I agreed upon was a "nonbias" or neutrality policy. I know I was extremely insistent upon it from the beginning, because neutrality has been a hobby-horse of mine for a very long time, and one of my guiding principles in writing "Sanger's Review." Neutrality, we agreed, required that articles should not represent any one point of view on controversial subjects, but instead fairly represent all sides.[61]

While Sanger and Wales agreed in principle at the outset, they have since expressed differences about the shift from the term *unbiased* to *neutral point of view*. At the start of Wikipedia, Sanger had ported Nupedia's "Avoid Bias" under Wikipedia's "Policies to Consider," but this policy was soon preempted/subsumed by Wales's "Neutral Point of View" article.[62] Sanger has since noted that he didn't approve of this shift as it causes confusion (e.g., using the expression "POV" as the opposite of "NPOV," when "biased" is preferable).[63] Not surprisingly, now that Sanger has started the encyclopedic project Citizendium, its "Neutrality Policy" favors the term *unbiased* over NPOV.[64] Yet before this recent difference about naming, at the outset of the Nupedia project Sanger and Wales were in agreement when challenged on the naíveté and/or impossibility of the policy. Sanger responded to the question of bias by invoking a principal that neutral contributions should lack ideological flavor:

> Nupedia aims to be as unbiased as possible; of course, some people will regard *this* as a political statement. We can't make everyone happy in this regard. In any event, we intend to represent all points of view, including those held by any significant minority of experts in a field, as fairly as possible. This would include creationists, Marxists, capitalists, and all manner of incendiary points of view. This should make for interesting reading at the very least. It should be added that Nupedia's contributors are expected to keep their own views in the background as much as possible. In other words, the point isn't merely to mention other views not favored by an article's author; it is to write in such a way that one cannot tell what view is favored by the article's author.[65]

The notion of not being able to tell the predilection of a contributor, a sort of ideological anonymity, is more fully developed in a corollary of NPOV on Wikipedia, "Writing for the Enemy":

> Writing for the enemy is the process of explaining another person's point of view as clearly and fairly as you can. The intent is to satisfy the adherents and advocates of that POV that you understand their claims and arguments. . . . Writing for the enemy contributes to the NPOV of Wikipedia. Wikipedians often must learn to sacrifice their own viewpoints to the greater good.[66]

For his part, Wales responded to someone troubled with the notion of unbiased by acknowledging the challenges and the importance of avoiding bias:

> Surely you will agree that there are _more_ or _less_ accurate, objective, fair, [un] biased ways of putting things. We should simply strive to eliminate all the problems that we can, and remain constantly open to sensible revisions. Will this be perfect? Of course not. But it is all we can do *and* it is the least we can do. . . . if you are trying to say that someone, somewhere will always accuse us of bias, I'm sure you're right. But we should nonetheless try our best to be objective. It doesn't strike me as particularly difficult. We will want to present a broad consensus of mainstream thought. . . . This does mean that sometimes we will be wrong! All the top scholars in some field will say X, but 50 years from now, we will know more, and X will seem a quaint and old-fashioned opinion. O.k., fine. But still, X is a respectable and valid opinion today, as it is formed in careful consideration of all the available evidence with the greatest care possible. That's the best we can do. And, as I say, that's also the least we can do.[67]

Consequently, this interest in unbiased, or at least less biased, claims about an understandable, or at least partially so, objective universe is central to Wikipedia collaborative culture. The notion of NPOV not only provides the epistemic foundation for the project, but also the intentional stance contributors should take while interacting. It makes it possible to "solve the problem of that jig-saw puzzle" for which H. G. Wells had hoped because, from this perspective, differing claims about the world can be fit together.

The Intersubjective Stance of Good Faith

In Wikipedia's collaborative culture, the scope of an open perspective includes not only the subject of collaboration, claims about the world, but also one's collaborators. In Wikipedia's "Writing for the Enemy" essay, one is encouraged to see things as others might:

> Note that writing for the enemy does not necessarily mean one believes the opposite of the "enemy" POV. The writer may be unsure what position he wants to take, or simply have no opinion on the matter. What matters is that you try to "walk a mile in their" shoes instead of judging them.[68]

The "Assume Good Faith" article on Meatball, where different communities discuss pan-wiki culture, characterizes this as "seeing others' humanity."[69] Indeed, one of the reasons Wikipedia's culture and practice are compelling to me is that it has influenced the way I approach controversy and conflict beyond Wikipedia; I have found these norms to be "a great way to end an argument in real life,"[70] which corresponds with scholars Yochai Benkler and Helen Nissenbaum's argument that while virtue may lead people to participate in such projects "participation may [also] give rise to virtue."[71] This sentiment and the challenges of collaborative culture are further reflected in Leuf and Cunningham's *The Wiki Way*: "People using Wiki bring their own preconceptions, agendas, and visions—like any community. The remarkable thing is how Wiki as community affects user interactions in an overall positive way."[72]

Unlike the relatively novel effect of NPOV on collaboration, Wikipedia is not the first online community to recognize the importance of, broadly speaking, good faith, and the challenges of other possibly competing values. In the Debian FOSS community, anthropologist Gabriella Coleman identifies a seeming paradox between liberal individualism/meritocracy and the community values of humility, detachment, generosity, and civility.[73] Similarly, Larry Wall, creator of the Perl programming language, playfully argues the success of his project is actually dependent on the coexistence of the seemingly contrary virtues of the individual programmer and the larger collaborative community. That is, programmers who exhibit the individual virtues of "laziness, impatience, and hubris," which often yield efficiency and quality, must also exhibit virtues of diligence, patience, and humility at the community level.[74] Leuf and Cunningham note that in wiki communities "participants are, by nature, a pedantic, ornery, and unreasonable bunch," yet "there's a camaraderie we seldom see outside our professional contacts."[75] Georg von Krogh, in his article on "Care in Knowledge Creation," identifies five dimensions relevant to the successful creation of knowledge within a community: mutual trust, active empathy, access to help, lenience in judgment, and courage.[76] Benkler and Nissenbaum argue that "commons-based peer-production" entails virtues that are both "self-regarding" (e.g., autonomy, independence, creativity) and "other-regarding" (e.g., generosity, altruism, camaraderie, cooperation, civic virtue).[77]

In subsequent chapters I too speak of seeming contradictions (e.g., benevolent dictators in egalitarian communities), but in the following sections I discuss good faith via four specific "virtues" or behaviors: assume the best, act with patience, act with civility, and try to maintain a sense of humor.

Assuming the Best of Others

Online communities often suffer the effects of Godwin's Law: as a discussion continues, someone is bound to make an unfavorable comparison to Hitler or Nazis. (Perhaps this is in part a consequence of the effects of computer-mediated communication, such as reduced social cues and anonymity, and the character of virtual community.[78]) A possible counteracting norm of this tendency is the guideline "Assume Good Faith." But before examining this norm in detail it is worthwhile to first note that good faith is associated with at least three collaborative wiki norms: good faith, "Assume Good Faith," and "Assume the Assumption of Good Faith."

Although present on Meatball, the wiki about wiki collaboration, the broad notion of good faith is not addressed by Wikipedia's guidelines; there is only a rather obtuse encyclopedic article adapted from the *Catholic Encyclopedia*'s legalistic treatment of the concepts of error and guilt.[79] But the notion of good faith does have colloquial usage, implicitly referring to a handful of concepts—much as I use it to signify the concepts of this section. This informal sense is captured in Meatball's description of good faith as a lack of intentional malice, an assumption that people are trying to do their best "for the greater good of the community," and friendliness, honesty, and caring.[80] The first two elements of this description are much the same, differing only in their subject: one's own positive intention and an assumption of others' positive intentions. It is on the latter assumption that Wikipedia focuses. The guideline of AGF is intended to counteract the common reflex to assume the worst of others, reminding us:

Well-meaning people make mistakes, and you should correct them when they do. You should not act like their mistake was deliberate. Correct, but don't scold. There will be people on Wikipedia with whom you disagree. Even if they're wrong, that doesn't mean they're trying to wreck the project. There will be some people with whom you find it hard to work. That doesn't mean they're trying to wreck the project either; it means they annoy you.[81]

Unlike unbiased view/NPOV, which was present at the start, "Assume Good Faith," in name, is a relatively new norm. The page was first created in March 2004; it received its first comment on its discussion page in February 2005.[82] (The first comment proposed "Assume Good Faith" become policy, although as previously noted AGF was demoted to a guideline in 2006 because it did not focus on behavior and was therefore difficult to enforce.) AGF's origins are most likely rooted in the "Staying Cool When the Editing Gets Hot" essay, which in October 2002 offered five "tips to consider when editing gets emotional," including avoid name calling and characterizing others' actions, take a breather if angry, ignore insults, and "assume the best about people."[83] "Assume the best" eventually found its way onto the "Etiquette" essay in January 2004,[84] but in August this was replaced with a link to the relatively new "Assume Good Faith" page.

While these norms of resisting name calling and of assuming the best seemingly arose in the context of everyday practice and in playground manners, even, they are also the subject of sociopsychological study. Under the fundamental attribution error, we often attribute the failures of others as evidence of a character flaw—but our own failings are construed as a circumstance of our environment.[85] That is, I succeed because of my genius and fail because of bad luck, whereas you succeed by chance and fail by your own faulty character. Not surprisingly, in a study of email collaboration Catherine Cramton found that in successful groups people typically give others the benefit of the doubt and make situational rather than categorical attributions about their behavior.[86] Less-successful groups included those that escalated hostility or were overly diplomatic—indicating the danger of both rancorous discord and facile consensus. From a psychological perspective, then, a cultural norm of assuming good faith can mitigate negative attributions.

AGF can also help set social expectations. This assumption is much like the popular aphorism "never attribute to malice what can be explained by stupidity."[87] The humorous Wikipedia essay "Assume Stupidity" notes that, "While assuming good faith is a fundamental principle on Wikipedia, it generally does not help you get over your anger at someone's, in your opinion, disturbing edits. Therefore, it is much more satisfying to also assume stupidity."[88] Fortunately, the official Wikipedia policy is more politic, as an assertion of stupidity might not be any more welcome than that of malice! Also, as the Meatball wiki cautions, low expectations can sometimes be damning: "Be warned that whatever we assume may become a self-fulfilling prophecy. We AssumeGoodFaith as a way of creating good faith, but assuming indifference or stupidity will encourage those modes as well." Yet at what point is the assumption of good faith exhausted? Meatball identifies a number of causes: some people might simply be trolling (being disruptive for their own fun), they might be an "angry [storm] cloud" (predisposed to conflict or having a bad day), they might be working at cross-purposes or be confused by a lack of transparency.[89] In fact, Wikipedia warns against ever attributing an editor's actions to bad faith "even if bad faith seems obvious"; one can always judge on the basis of behavior rather than assumed intentions.[90] For example, the invocation of "Assume Good Faith," because it is about intentions, can become an act of bad faith itself, leading to the awkwardly named exhortation to "Assume the Assumption of Good Faith":

> In heated debates, users often cite AGF. However, the very act of citing AGF assumes that the opponent is assuming bad faith. Carbonite's law tells us, "the more a given user invokes 'Assume Good Faith' as a defense, the lower the probability that said user was acting in good faith."[91]

To this end, the AGF guideline wisely recommends, "If you expect people to 'Assume Good Faith' from you, make sure you demonstrate it. Don't put the burden on others. Yelling 'Assume Good Faith' at people does not excuse you from explaining your actions, and making a habit of it will convince people that you're acting in bad faith."[92]

However, an assumption that counters cognitive bias and sets social expectations still stops short of coming to know and understand others. Here the norm of "WikiLove," "a general spirit of collegiality and mutual understanding,"[93] makes the same sort of connection that I am attempting to make in this chapter: an open perspective (or love) of knowledge melded with caring attitude (or love) toward others. Or as Wales said in his 2004 "Letter": "The only way we can coordinate our efforts in an efficient manner to achieve the goals we have set for ourselves, is to love our work and to love each other, even when we disagree."[94] This is most clearly reflected in a prominent Wikipedian's declaration that Wikilove is the most important principle of all:

I believe that we need to highlight the mission of providing a great, free encyclopedia, along with the core principle _how_ we want to accomplish it. And the single most important principle I can think of here is not "anyone can edit." It's not even NPOV or any other policy. It's "WikiLove"—of which our commitment to openness is only an expression. We share a love of knowledge, and we treat everyone who shares the same love with respect and goodwill. (That's the idea, at least.)[95]

At this point I want to point out a possible transition between "Assume Good Faith" and "WikiLove." In the wide range of literature on interacting with others one might discern three not necessarily exclusive ways of orientating toward others: self, selfless, and group.[96] The first might be characterized as the strategic choice of a "rational egoist." Whereas perspective taking often yields "joint gains," this does not preclude it from being a self-interested behavior that mitigates the erroneous attributions and impasses that impairs one's own interests.[97] For example, it is in the self-interest of a negotiator to "understand" the perspective (e.g., the best alternative to negotiated agreement) of her opponent. Another approach is at the other extreme. Here, some actions are construed as being selflessly "other" orientated, even when counter to self- or group interests. This may be present in particular types of dialogue, empathy, and caring.[98]

Another common focus is on the group. In the literature of political economy, "collective action" refers to circumstances in which cooperation is beneficial to the group, and each member, but only if others cooperate as well. In such situations prosocial norms—and a willingness to punish defectors—can support sustained cooperation.[99] Obviously, the importance of trust, empathy, and reciprocity on building community relationships and facilitating the exchange of ideas is key.[100] Trust is characterized by group members who are honest in negotiating commitments, who make "a good faith effort" to abide by their explicit—and implicit—commitments, and don't take excessive advantage of others even when opportunities to do so arise.[101] Furthermore, trust not only affects the expectations of an interaction, but also the construal of it afterward.[102] Indeed, in "good faith" interactions, trust is the supposition that even though one disagrees and hasn't been able to see and understand from another's perspective, one might be missing something. For example, in his study of consensus-based decision making within the Society of Friends, Michael Sheeran notes that a dissenting Quaker might respond, "I disagree but do not wish to stand in the way" because: "For religious reasons, a person may prefer the judgment of the group as 'sincere seekers after the divine leading' to that person's individual judgment. In more secular terms, an individual may recognize the possibility that everyone else is right."[103] Trust in others implies a sense of humility toward one's self as noted in Kizor's Law of Humility: "Better an editor who's often wrong and knows it than an editor who's very seldom wrong and knows it."[104]

All that said, the debate of whether all altruism is necessarily "egoistic" is a complex one, but Wikipedia might serve as a relevant case for those interested in the discussion.[105] (Obviously, anonymous contribution is a provocative topic for those concerned with the motives of seemingly altruistic contributors.) And, in the case of Wikipedia, one might ask this more specific question: To what extent is good faith simply a matter of being a more effective and respected Wikipedian, a matter of group altruism, or something more? I would characterize the text on and discussion related to good faith as predominately oriented toward the group. This does not preclude egoistic self-satisfaction, or a transcendent intention, but Wikipedia discourse is rooted in extending good faith and WikiLove in service of a mutual love of knowledge: "We are all here for one reason: we love accumulating, ordering, structuring, and making freely available what knowledge we have in the form of an encyclopedia of unprecedented size."[106]

Patience

A deficient collaborative culture might be characterized as temperamental and brittle because participants are uneasy and defensive; and existing structures and agreements easily fracture, providing little common ground and means for facilitating agreement. Its opposite, a well-working collaborative culture, might be characterized by patience as participants do not easily panic or escalate conflict. As a Wikipedia essay counsels: "The world will not end tomorrow."[107]

In response to community concerns and conflict generated in response to Wikipedia office actions, where the Foundation office removes "questionable or illegal" content given complaints including "defamation, privacy violations or copyright infringement,"[108] Jimmy Wales responded that in such circumstances the community should: "Assume Good Faith. It could be a mistake, it could be a poor decision, it could be a very strange emergency having to do with a suicide attempt. . . . In general, there is plenty of time to stop and ask questions."[109]

Another source of contention is the many differing positions about what kind of encyclopedia Wikipedia should be. Should it address topics like those of any other encyclopedia, or is there also room for encyclopedic articles about every episode of *Buffy the Vampire Slayer*? On this question of scope, there is a range of philosophical views (i.e., "isms").[110] For example, there is deletionism (rigorous criteria for a uniformly worthwhile article must be met, otherwise delete), mergism (merge challenged information into an existing article rather than have it stand alone), essentialism (include traditionally nonencyclopedic information but only if it is notable and verifiable), and inclusionism (keep as long as an article has some merit). And yes, at present, every one of the 144 episodes of *Buffy* does have its own article.[111]

Perhaps an explanation of Godwin's Law is that, as discussed, participants come to believe that the issue at hand is eclipsed by larger, more abstract matters, a conflict of principles, a battle between good and evil. As the essay "Don't Escalate" notes, "we need to watch how many layers of indirection we're piling onto the discussion and try not to stray too far from the substantive issue."[112] The recourse of patience can mitigate such escalation: "Cease what you are doing. Count to 10. Take a break. Read a book. Watch some videos on Youtube. Don't edit. Don't press the 'save page' button. Do what you have to do to cool down."[113] Consider a discussion as to whether the contentious "Articles for Deletion" process could be suspended for a month,[114] in which a Wikipedian recommended that instead of panicking,

> both camps could Assume Good Faith and relax a bit, each not thinking that the "other guys" are a bunch of deranged encyclopedia-haters who want to destroy everything in an orgy of deletion and/or garage band stubs [incomplete vanity articles].:) A lot of people are currently disagreeing over what sorts of articles merit inclusion in Wikipedia, but it's not like most of those people think Wikipedia's going to go down in flames if the "wrong" standards are picked. At least, they shouldn't. Wikipedia is more resistant than that.[115]

Patience is further implicated by "Assume Good Faith," since frustrating behavior resulting from ignorance, rather than malice, is remedied in time, as the "Please Don't Bite the Newcomers" guideline cautions:

> New contributors are prospective "members" and are therefore our most valuable resource. We must treat newcomers with kindness and patience—nothing scares potentially valuable contributors away faster than hostility. It is impossible for a new-

comer to be completely familiar with the policies, standards, style, and community of Wikipedia (or of a certain topic) before they start editing. If any newcomer got all those things right, it would be by complete chance.[116]

And the guideline of "Do Not Disrupt Wikipedia to Illustrate a Point" has a similar concern with dampening an escalation toward principle and returning to the immediate concern at hand,[117] as does the essay "Wikipedia Is Not Therapy":

> Wikipedia is not therapy. If a user has behavior problems which result in disruption of the collective work of creating a useful reference, then their participation in Wikipedia may be restricted or banned entirely. This should not be done without patiently discussing any problems with the user, but if the behavior is not controlled, ultimately the project will be protected by restricting the user's participation in the project.[118]

Finally, the technology of wiki itself furthers patience as a change can always be reversed without fear of permanent damage; as software developer and author Karl Fogel notes with respect to producing free and open source software: "version control means you can relax."[119]

The extent to which patience is extended to problematic participants has been a source of (pleasant?) surprise for Wikipedia cofounder Jimmy Wales, who once noted, "when I am asked to look into cases of 'admin abuse' and I choose to do so, I generally find myself astounded at how nice we are to complete maniacs, and for how long."[120] Yet such patience can be exhausted, as noted by Larry Sanger, the other Wikipedia cofounder and present apostate:

> A second school of thought held that all Wikipedia contributors, even the most difficult, should be treated respectfully and with so-called WikiLove. Hence trolls were not to be identified as such (since "troll" is a term of abuse), and were to be removed from the project only after a long (and painful) public discussion.[121]

Not surprisingly, the balance of patience to be extended continues to be a topic of discussion. Yet there are cases in which participants disappoint all good assumptions, wear patience thin, and remain lovable only to their mothers; up to, and even after, this point, participants are still expected to remain civil.

Civility

A subtle, but important, incoherence is found within the Wikipedia "Policies and Guidelines" page: "Respect other contributors. Wikipedia contributors come from many different countries and cultures, and have widely different views. Treating others with respect is key to collaborating effectively in building an encyclopedia."[122] Are Wikipedians to genuinely respect all others, or (merely) treat them with respect? A comment in the "Civility" policy points to the second interpretation: "We cannot always expect people to love, honor, obey, or even respect another. But we have every right to demand civility."[123] I make this distinction between genuine respect and acting with respect based on Mark Kingwell's useful definition of civility in public discourse:

> It is true that civility as I characterize it is related to mutual respect, but there is a crucial difference: genuine respect is too strong a value to demand . . . in a deeply pluralistic society. The relative advantage of civility is that it does not ask participants to do anything more than treat political interlocutors as if they were worthy of respect and understanding, keeping their private thoughts to themselves.[124]

Consequently, civility acts as both a baseline for building a culture of good faith and as a last line of defense against escalation. Despite expectations to act in good faith, "Assume Good Faith," walk in another's shoes, see another's humanity, and love and respect one another, failing all of this, Wikipedians should still be civil and treat each other with respect. This means refraining from "personal attacks, rudeness, and aggressive behaviours that disrupt the project and lead to unproductive stress and conflict."[125] Otherwise, as Kingwell notes, "when civility fails, we all lose, because as citizens we lose the possibility of justice, and of a genuinely shared political community."[126] Or, as Wikipedia warns: "Being rude, insensitive or petty makes people upset and prevents Wikipedia from working properly."[127] A lack of civility is self-reciprocating, in that alienation begets alienation, and other faults, such as hypocrisy, soon follow, which "has the same effect on good faith that termites have on wooden houses."[128]

Aside from the communicative aspect of dampening counterproductive hostility, historically, civility has also played a role in the production, or at the least legitimation, of knowledge. In *A Social History of Truth*, Steven Shapin notes that "gentlemen," as signified in part by their civility, were thought of as arbiters of truth because their privileged status allegedly rendered them immune from external pressure: the man who did not have to labor for his bread was least likely to "shift" his views.[129] (Though one might argue that the gentleman's privileged status certainly biased his perspective.) Although civility is still important within Wikipedia, it is not relied on as a premodern performance to represent social standing and consequently the ability to legitimate knowledge. Rather, encyclopedic knowledge emerges from civil discourse between people who may be strangers; civility facilitates the generation of knowledge rather than being a proxy for social standing or institutional affiliation.[130]

That said, civility can be a difficult principle for the community, as people vary in their outspokenness and sensitivity. In the summer of 2009 a large poll was conducted among English Wikipedians, asking whether the civility policy was satisfactory, or abused or selectively enforced (i.e., were people baited and then attacked with this policy); also, was its application consistent across all parts of Wikipedia, and did it interfere with clarity? The resulting summary concluded that:

> The majority of people feel the current civility policy is too lenient, and that it is inconsistently applied and unenforceable. Most people feel that civil behaviour applies as much on personal talkpages as elsewhere, and that there are particular problems with civil behaviour on Recent Changes Patrol and Admin Noticeboards. Almost everyone feels we are too harsh on new users, though just over half the people feel that when it comes to experienced users that expectations of behaviour depends on context and the people involved. Most people feel that baiting is under-recognised, although it was noted that it is difficult to recognise baiting, and that people have a choice in how they respond.[131]

This does not imply civility will be abandoned as a policy; the principle at least will persist, although it and its implementation will continue to be discussed, no doubt.

Humor

Humor is not a policy or guideline of Wikipedia, but it suffuses the culture and is the true last resort when faced with maddening circumstances.[132] Certainly, Wikipedia is the butt of many jokes. The satirical newspaper the *The Onion* has made fun of the often-contentious character of Wikipedia with an article about the U.S. Congress abandoning an attempt at a wiki version of the Constitution; it also lampooned Wikipedia's reliability with the article "Wikipedia Celebrates 750 Years of American Independence."[133] Wikipedia has also been the source of fun for many Web comics, such as a *Penny Arcade* strip entitled "I Have The Power," showing the evil cartoon character Skeletor changing He-man's description from "the most powerful man on earth" to "actually a tremendous jackass and not really that powerful."[134] Wikipedians are also capable of laughing at themselves. In August

2009 there were over seven hundred articles listed in Wikipedia's humor category,[135] including a dozen or so songs and poems, such as "Hotel Wikipedia" and "If I Were an Admin."[136] An excerpt from my favorite, based on a ditty from Gilbert and Sullivan's *The Pirates of Penzance*, best captures the character of Wikipedians:

> I am the very model of a modern Wikipedian, / My knowledge of things trivial is way above the median, / I know, and care, what Kelly Clarkson's next CD might just be called, / And all the insults Hilary and Lindsay to each other bawled. / I'm very well acquainted, too, with memes upon the Internet, / I think the dancing hamster would be excellent as a pet. / About the crackpots' physics I am teeming with a lot o' news, / The Time Cube has but four sides and it's not got a hypotenuse.[137]

Nor is humor relegated only to the funny category. It is present in many of the norms discussed so far, capturing the difficult character of these principles and their practice. For example, the "In Bad Faith" essay collects examples of bad faith, such as "If I compromise, they'll know it's a sign of weakness," and "That policy page is wrong, because it doesn't describe what I do. I'll fix it."[138] The "Neutral Point of View" policy notes that when you are writing for the enemy "the other side might very well find your attempts to characterize their views substandard, but it's the thought that counts."[139] The "Don't Be Dense" essay asks the reader to remember that "'Assume Good Faith' is a nicer restatement of 'Never assume malice when stupidity will suffice.' Try not to be stupid either."[140] In recognition of the unavoidable absurdity of "isms" there is the most absurd, though quite reasonable, philosophy of all, the AWWDMBJAWGCAWAIFDSPBATDMTD faction: "The Association of Wikipedians Who Dislike Making Broad Judgements About the Worthiness of a General Category of Article, and Who Are In Favor of the Deletion of Some Particularly Bad Articles, but That Doesn't Mean They are Deletionist."[141] And given the hundreds of user-created Laws of Wikipedia, Kmarinas86's Law of Contradiction recommends that "When one law contradicts the other(s), the funniest one applies first."[142]

Humor serves as an instrument of anxiety-releasing self-reflection. As the saying goes, if you can't laugh at yourself, who can you laugh at? Michael Schrage, author of the 1990 book *Shared Minds: The New Technologies of Collaboration,* alludes to the importance of humor when he writes:

> Designing for collaboration requires an architect with a sense of humor. After all, collaborative relationships have to cope with the misunderstandings as well as the epiphanies, and the tool should be able to support them all with grace. Creating an

environment that stimulates the relaxed intensity that marks effective collaboration is a craft, not a science. It requires both an aesthetic sense and a grasp of functionality.[143]

Humor is also an instance of intellectual joy, like the many jokes and puns common to geek culture. Ultimately, Wikipedia is supposed to be enjoyable. When circumstances arise such as battling spammers, trying to discern the well-meaning newbie from a troll, politicking over the deletion of an article, and other inherently contentious and non-fun activities, humor serves as a way to restore balance. At times, it may also disrupt balance. For example, sarcasm is a brand of frequently unproductive humor, as parodied in the "Sarcasm Is Really Helpful" essay: "Sarcasm works really well in online media, because it's so easy to pick up on without all of those pesky extratextual cues. It's hard to see how the employment of sarcasm could possibly be counterproductive."[144] Also, many Wikipedians dread April 1 because this tomfoolery isn't present and understood in all cultures, some use the date as an excuse for outright vandalism, and many object to any change of encyclopedic articles for humorous purposes. English Wikipedia currently solves this problem every year by featuring a new article on a topic so odd you would think it is a prank, but it is not.[145] Sometimes the values of civility and humor are posed as opposites:

P.S. I know I'm not alone in saying that I have considered leaving Wikipedia on several occasions not because of incivility or personal attacks, but because there are people who can't and refuse to take an obvious joke. The humorless people will ruin Wikipedia before those who aren't prim, proper and civil.[146]

However, I find that gentle humor and civility more often than not are complementary. When they are not, the question often comes down to—just as it may in the schoolyard—who is the butt of the joke.

Conclusion

Wikis are a relatively novel way of working together: online, asynchronous, possibly anonymous, incremental, and cumulative. Do these features alone explain the success of Wikipedia? Not quite. Each also has possible demerits. Flame-ridden, scattered, unaccountable, half-baked piles of bunk are a possible future for any wiki. As the WikiLove essay notes, "Because people coming from radically different perspectives work on Wikipedia together—religious fundamentalists and secular humanists, conservatives and socialists, etc.—it is easy for discussions to degenerate into flamewars."[147]

So, in addition to technology, a community's collaborative culture is an important factor in determining what its future holds. Wiki communities are also a fascinating subject of study because one can closely follow the emergence of and discourse on their culture: what is important, what is acceptable, and what does it all mean? On a wiki, the regenerative, recursive, or dual nature of community policy and practice renders discussions about these questions intensely transparent—not that this makes it necessarily easy to filter and understand. As Leuf and Cunningham wrote in 2001, "Wiki culture, like many other social experiments, is interesting, exciting, involving, evolving, and ultimately not always very well understood."[148]

In the case of the English Wikipedia, there is a collaborative culture that asks its participants to assume two postures: a stance of neutral point of view on matters of knowledge, and a stance of good faith toward one's fellow contributors. Whereas NPOV renders the subject matter of a collaborative encyclopedia compatible, good faith makes it possible to work together. It is as if the NPOV permits collaborators to bring together the "scattered and ineffective mental wealth" of H. G. Wells's jigsaw. However, this doesn't mean the process of working together will be effective or enjoyable. Therefore, a culture of assuming the best of others, and demonstrating patience, civility, and humor facilitates collaborating with one's peers, of varied persuasions, to fit the pieces together. As the "Collaboration First" essay declares: "A productive contributor who cannot collaborate is not a productive contributor."[149]

4 The Puzzle of Openness

Problematic users will drive good users away from Wikipedia far more often than good users will drive away problematic ones.
—Extreme Unction's Third Law

Trolls are the driving force of Wikipedia. The worst trolls often spur the best editors into creating a brilliant article with watertight references where without the trollish escapades we would only have a brief stub.
—Bachmann's Law

A central aspiration in the pursuit of a universal encyclopedia is increased access to information: an opening of opportunity and capability to anyone with a desire to learn. Ironically, such an encyclopedia only became possible with universal access to its production. However, Wikipedia's openness, based on the inspiration of the Free and Open Source Software (FOSS) movement and the capabilities of hypertext, is not a collaborative panacea. The two, at odds, "laws" of Wikipedia that begin this chapter are evidence that openness has advantages and disadvantages—and people don't even agree about which is which. (Since neither law is funny, Kmarinas86's Law of Contradiction, in which the most humorous wins, is of little help.) In fact, like the issues of consensus and leadership addressed in the next two chapters, openness, including to those who may alienate good users or drive them to brilliance, is a bit of a puzzle itself.

Wikipedia's claim of openness is seen in its motto: "Wikipedia, the free encyclopedia that anyone can edit."[1] But what do the terms *openness*—and *anyone*—actually mean? Because of the ascendancy of FOSS, *open* is now a buzzword, becoming a prefix to even such well-established notions as democracy and religion.[2] Additionally, when contemporary sources speak of openness as an attribute of community, it is often in an overly simplistic way; projects like the Linux kernel, Apache Web server, and Wikipedia are often mischaracterized by way of an inappropriate, if not naïve, extreme. A utopian rendering of openness is that "anything goes": there are no community structures or norms, anyone can do anything they please.

This understanding of "anything goes" is untenable: some level of structure is inevitable in social relations, and often necessary to support other values. In his 1911 book *Political Parties,* Rober Michels wrote of the development of an oligarchy within democratic parties as an "Iron Law." In 1970 Jo Freeman wrote about the "tyranny" present in seemingly egalitarian feminist groups of the earlier decade: "'Structurelessness' is organisationally impossible. We cannot decide whether to have a structured or structureless group; only whether or not to have a formally structured one." And more recently, Mitch Kapor expressed a similar sentiment with respect to the early management of the Internet when he noted: "Inside every working anarchy, there's an Old Boy Network."[3]

To be fair, Wikipedia has not helped this confusion given its early rule of "Ignore All Rules." Granted, it is clever to have a rule dismissing rules, and its substance is of merit: recognizing the robustness of wikis and painfulness of bureaucracy: "If a rule prevents you from improving or maintaining Wikipedia, ignore it."[4] However, such a bald, even if humorous, assertion is bound to require qualification. The essay "What 'Ignore All Rules' Means" explains that novices should feel free to contribute, don't be overly legalistic but work in the spirit of improving the encyclopedia, and there is no substitute for good judgment. It doesn't mean "every action is justifiable," nor is it an excuse or exemption from accountability.[5]

Yet, these difficulties do not mean the notion of openness should be jettisoned altogether. The prevalence of the term *open* in contemporary discourse is indicative of something important, and provides a window into understanding the English Wikipedia community. For example, in 2001 Jimmy Wales posted a "Statement of Principles," the first of which is that "Wikipedia's success to date is entirely a function of our open community. This community will continue to live and breathe and grow only so long as those of us who participate in it continue to Do The Right Thing."[6] But what constitutes "the right thing" in the context of openness?

In the following sections, I portray Wikipedia in light of five characteristics of what I call an *open content community*. This concept permits me to distinguish between a type of content, such as FOSS, and the community that produces it.[7] (A "closed" company can produce software under an "open" license.) It also permits me to identify more specific values implicit in discussions about openness, such as transparency and nondiscrimination. Openness and its related values are then considered in light of four cases in which we see the Wikipedia community wrestling with how it conceives of itself. In the first case I return to the question of whether Wikipedia is really something "anyone can edit"? That is, when Wikipedia implemented new technical features to help limit vandalization of the site, did it make Wikipedia more or less open? In the second case I describe the way in which a maturing open content community's requirement to interact with the world beyond Wikipedia affects its openness. In this case, I review Wikipedia's "office action" in which agents of Wikipedia act privately so as to mitigate potential legal problems, though this is contrary to the community values of deliberation and transparency. Third, I briefly review concerns of how bureaucratization within Wikipedia might threaten openness. Finally, I explore a case in which a closed (female-only) group is set up outside of, and perhaps because of, the "openness" of the larger Wikipedia community.

Open Content Communities

I use the word *community* to speak of a group of interdependent people who "participate together in discussion and decision making and who share certain practices that both define the community and are restored by it."[8] Wikipedia community members do share common practices and norms; as we've seen, they share a collaborative culture. Furthermore, the Wikipedia community can be further understood as "prosocial" in that it exhibits behavior that is intentional, voluntary, and of benefit to others.[9] But even if we can defensibly claim it is a prosocial community, can anyone claim that it is truly open? Such a question requires a better sense of what *open* means. After reviewing its many uses as inspired by FOSS, I characterize openness in this context as an *accessible and flexible* type of *collaboration* whose result may be *widely shared.*[10] More specifically, an *open content community* is characterized by:

- Open content: provides content that is available under FOSS licenses.
- Transparency: makes its processes, rules, determinations, and their rationales available.

• Integrity: ensures the integrity of the processes and the participants' contributions.

• Nondiscrimination: prohibits arbitrary discrimination against persons, groups, or characteristics not relevant to the community's scope of activity. Persons and proposals should be judged on their merits. Leadership should be based on meritocratic or representative processes.

• Noninterference: the linchpin of openness, if a constituency disagrees with the implementation of the previous three values, the open content license permits the constituents to take the content and commence work under their own conceptualization without interference. While "forking" is often complained about in open communities—it can create some redundancy/inefficiency—it is an essential characteristic and major benefit of open communities as well.

Although the first and last characteristics provide a "bright line" with which one can distinguish between instances of open content by their copyright licenses and the consequent ability to fork, the social values of transparency, integrity, and nondiscrimination do not provide an equally clear demarcation. (What counts as open or free content has not always been an easy question either, but we do now have the "Free Software Definition," "Open Source Definition," and "Definition of Free Cultural Works."[11]) Additionally, although the often-voluntary character of contribution is not directly related to openness, it is critical to understanding the moral/ideological light in which many members view their participation. Each of these characteristics is explored so as to identify the context and values inherent in discussions by the community about itself.

Open Content

As noted, what is often meant by the term *open* is a generalization from the FOSS movement. Communities marshaling themselves under this banner cooperatively produce, in public view, software, technical standards, or other content that is intended to be widely shared. Fortunately, there are now a number of excellent works on the FOSS phenomenon;[12] therefore, I only provide a brief description to clarify what is meant by "open content" and to identify one of Wikipedia's main inspirations.

The free software movement was spearheaded by Richard Stallman at MIT in the 1980s. When Stallman found it difficult to obtain the source code of a troublesome Xerox printer, he feared that the freedom to tinker with and improve technology was being challenged by a different, proprietary conceptualization of information.[13] To respond to this shift he created two organizations: the GNU Project in 1984, which develops and maintains free software, and the Free Software Foundation (FSF) in 1985,[14] which houses legal and advocacy efforts. Perhaps most importantly he wrote the first version of the GNU General Public License (GPL) in 1989. The GPL is the seminal copyright license for "free software"; it ensures that the "freedom" associated with being able to access and modify software is maintained with the original software and its derivations. It has important safeguards, including its famous reciprocal provision: if you modify and distribute software obtained under the GPL license, your derivation must also be licensed and available under the GPL. (This provision is sometimes referred to as "viral" though some find this label derogatory.) Because such software is often of little or no cost to acquire, cost and freedom are sometimes conflated; this is answered with the slogan "Think free as in free speech, not free beer."

In 1991, Linus Torvalds started development of Linux, a UNIX-like operating system kernel, the core computer program that mediates between applications and the underlying hardware. While it was not part of the GNU Project, and differed in design philosophy from the GNU's kernel (named "Hurd"), it was released under the GPL. While Stallman's stance on "freedom" is more ideological, Torvalds's approach is more pragmatic. Furthermore, other projects, such as the Apache Web server, and eventually Netscape's Mozilla Web browser, were developed under similar open licenses except that, unlike the GPL, they often permit proprietary derivations. With such a license, a company may take open source software, change it, and include it in the company's product without releasing its changes back to the community.

The tension between the ideology of free software and its other, additional benefits led to the concept of open source in 1998. The Open Source Initiative (OSI) was founded when Netscape was considering the release of its browser as free software; during these discussions, participants "decided it was time to dump the moralizing and confrontational attitude that had been associated with 'free software' in the past and sell the idea strictly on the same pragmatic, business-case grounds that had motivated Netscape. They brainstormed about tactics and a new label. 'Open source,' contributed by Chris Peterson, was the best thing they came up with."[15] Under the open source banner the language and ideology of freedom were sidelined so as to highlight pragmatic benefits and increase corporate involvement.

The benefits of openness are not limited to software. Because the FSF felt the documentation that accompanies free software should also be free it created the GNU Free Documentation License (GFDL) in 1999. Of course, in the new millennium this model of openness has extended to forms of cultural production beyond technical content. Wikipedia's cofounder Jimmy Wales has stated that a seminal article by Eric Raymond, which likened FOSS production to that of a vibrant, decentralized bazaar, "opened my eyes to the possibility of mass collaboration."[16] In fact, in October 2001, when Wikipedia was not even one year old, Wales collected those principles he thought were responsible for and would continue to be needed for Wikipedia's success. In his "Statement of Principles," Wales wrote that "success to date is entirely a function of our open community." As Nupedia and Wikipedia were licensed under the GFDL from the start, "The GNU FDL [GFDL] license, the openness and viral nature of it, are fundamental to the long-term success of the site."[17]

Ironically, Wikipedia is not looked upon favorably by some prominent FOSS developers. Eric Raymond has characterized Wikipedia as a "disaster" that is "infested with moonbats";[18] in his view Wikipedia is an unsuitable case of the open source model because the merit of software developers and their code can be judged by objective standards (e.g., speed or efficiency), but knowledge claims cannot. A participant on the geek discussion site Kuro5hin writes, "People love to compare Wikipedia to Open Source but guess what: bad, incorrect code doesn't compile. Bad, incorrect information on the 'net lives on and non-experts hardly ever notice the mistake."[19] FOSS scholar Felix Stalder notes this difference between functional and expressive content is one of the key differences between "open source" and "open culture."[20]

In time, because the GFDL was intended to accompany the textual documentation of software, and was perceived by some as not being flexible enough, new nonsoftware content licenses have appeared. More widely, the Creative Commons project, launched in 2001, provides licenses for the sharing of texts, photos, and music. Law professor Lawrence Lessig, a founder of Creative Commons, helped popularize the notion of freedom and openness in domains beyond software with his book *Free Culture*.[21]

Wikipedia is probably the best-known example in the wider free culture movement today.

Transparency and Integrity

On the Meatball wiki, "a common space for wiki developers and proprietors from all over the Internet to collaborate," the values of transparency and integrity are partially captured by what it calls "Fair Process," which itself includes the three principles of engagement, explanation, and clarity; fair process is particularly important in voluntary communities, where "because fair process builds trust and commitment, people will go above and beyond the call of duty."[22]

While in the past some have warned of "eroding accountability in computerized societies," more recently others have argued that the open development of FOSS may be an exception, and even provide a model for achieving accountability for other technologies or institutions.[23] Consequently it shouldn't be surprising that transparency has come to be an attributed feature of Wikipedia. Jill Coffin explains that transparency "allows participants to understand the reasoning behind decisions, contributing to trust in the Wikipedia process. It also allows newbies a means to understand informal community protocol and culture, as well as reduce abusive practice." Wiki technology and culture promote the documentation of proposals, discussions, and decisions—everything, actually. Integrity can then flow from the accountability inherent to such transparency: the record is there for all to see. Coffin relates this to a famous Linux aphorism: "Schlock and chaos are avoided due to the watchful eyes of the many, exemplifying Linus' Law, coined and articulated by hacker Eric Raymond as 'Given enough eyes, all bugs are shallow.'"[24] Some scholars even argue that Wikipedia "embodies an approximation" of philosopher Jürgen Habermas's notion of "rational discourse" (i.e., noncoercive, open to participation, and discursive).[25] At Wikipedia, the importance and hoped-for effects of transparency can be seen in the expectations of its "stewards," who have significant power in administrating all other user rights:

> Steward activity is visible in the Meta rights log. When a request is fulfilled, stewards should note what they did at the local request page (each new request should be accompanied by a link to this) or on the Meta request page. Steward discussions should occur on Meta, rather than by e-mail, so people can understand the stewards' decisions and ways of working.[26]

However, just as a naïve rendering of openness as "anything goes" is overly simplistic, so is the sense that just because something has been posted on the Web one has achieved a perfect level of transparency and integrity. As we shall see, there might still be other private communication channels and sometimes too much information can be as disabling as not having any at all.

Nondiscrimination

A common tendency in groups is to adopt a parochial in-group/out-group mentality; Wikipedia cultural norms attempt to counter this. In the 2001 "Statement of Principles" Wales wrote, "Newcomers are always to be welcomed. There must be no cabal, there must be no elites, there must be no hierarchy or structure which gets in the way of this openness to newcomers." This is further reflected in the famous Wikipedia maxim "Please Do Not Bite the Newcomers."[27]

Beyond newcomers, there are also norms of nondiscrimination with respect to behavior and beliefs. In the wikiEN-l thread entitled "Wikipedia and autism," Tony Sidaway wrote of the treatment of two admittedly difficult contributors: "Both of them have expressed a strong wish to produce work for Wikipedia. Both of them produce articles that appear weird to non-autists. In my opinion, neither represents a threat to Wikipedia commensurate to the treatment they have received."[28] Wikipedians then discussed how they might best work with and encourage such contributors. Also, as seen in the neo-Nazi scenario at the beginning of this book, Wikipedia administrator MattCrypto unblocked a "racialist" because he thought it was unfair to block individuals because of their affiliations rather than for their actions on Wikipedia. Even those with criticisms of Wikipedia should be welcomed if they connect in a constructive way, as Wales notes in his "Statement of Principles":

> Anyone with a complaint should be treated with the utmost respect and dignity. They should be encouraged constantly to present their problems in a constructive way in the open forum of the mailing list. Anyone who just complains without foundation, refusing to join the discussion, I am afraid I must simply reject and ignore. Consensus is a partnership between interested parties working positively for a common goal. I must not let the "squeaky wheel" be greased just for being a jerk.[29]

However, it is interesting to note that the "Statement of Principles" of October 27, which is a seminal articulation of the Wikipedia ethos, appeared after two other messages relevant to Wikipedia openness. Jimmy Wales and Larry Sanger are often inappropriately placed at extremes of the "crowds versus experts" continuum; however, Sanger has welcomed mass participation under the guidance of experts and Wales has recognized the challenge of mass participation as Wikipedia continues to grow, as seen in a controversial message Wales posted on October 18:

> One of the wonderful things about the wiki software, and something that has served us very well so far, is that it is totally wide open. I suspect that any significant deviation from that would kill the magic of the process.
>
> On the other hand, we really are moving into uncharted territory. Wikipedia is already, as far as I know, the most active and heavily trafficked wiki to ever exist. It seems a virtual certainty that the wide open model will start to show some strain (primarily from vandalism) as we move forward.
>
> I have this idea that there should be in the software some concept of "old timer" or "karma points." This would empower some shadowy mysterious elite group of us to do things that might not be possible for newbies. Editing the homepage for example. We already had one instance of very ugly graffiti posted there (a pornographic cartoon).[30]

This message was a faux pas on Wales's part. In chapter 6, I note that an open content community is often led by a "benevolent dictator," and the community deals with the anxiety arising from the tension between the egalitarian ethos and autocratic leadership by way of irony and humor. In this message Wales speaks of being a dictator and of a cabal in much the same way—without appreciating that the joke doesn't work when he tells it. One week later Wales was forced to explain:

> In a letter to wikipedia-l, I injudiciously used the word "cabal" and made reference to a "shadowy mysterious elite." This was a very poor choice of words on my part. I thought that many or most people would understand it for what it was—the notion of a non-existent cabal, allegedly controlling things, when in fact there is not one, would be well understood.
>
> Let me be clear. In wikipedia, there should be no elites. All legitimate participants, no matter how much they may disagree on political, philosophical, or other issues, should always be able to edit pages in the same fashion as they can now. Only behavior that truely and clearly rises to the level of vandalism should be fought with extremely cautious uses of software security measures.[31]

And the following day Wales posted his "Statement of Principles" on the wiki, further highlighting the importance of openness to Wikipedia's success. Even so, fears of a cabal continue to arise every so often; it is human nature and a social inevitability for practice to sometimes fall short of principle and for people to be suspicious of those in power. On the one hand, Wikipedians frequently raise concerns about transparency, integrity, and discrimination;[32] on the other, Extreme Unction's First Law notes that "if enough people act independently towards the same goal, the end result is indistinguishable from a conspiracy."[33]

Noninterference

Simply, if the content is available under an open/free license, those dissatisfied with it or the community can take the content and work on it elsewhere.[34] Steven Weber notes the importance of forking by claiming: "The core freedom in free software is precisely and explicitly the right to fork." While I don't consider it to be the "core freedom" but rather a critical social *implication* of open/free content, it is a "fundamental characteristic" of FOSS. And I do agree that "to explain the open source process is, in large part, to explain why [forking] does not happen very often and why it does when it does, as well as what that means for cooperation."[35] To this end, David Wheeler likens forking to "the ability to call for a vote of no confidence or a labor strike . . . . Fundamentally, the ability to create a fork forces project leaders to pay attention to their constituencies."[36]

Forks of Wikipedia content have happened and continued to be threatened and discussed. For example, because of a misunderstanding about the possibility of Wikipedia carrying advertising, the Spanish-language Wikipedia was forked into *Enciclopedia Libre Universal*.[37] (The misunderstanding has since been resolved and Spanish Wikipedia has superseded the fork.) Or, as noted, Larry Sanger's dissatisfaction with the lack of respect for expert contributors at Wikipedia led him to start the Citizendium project, which uses the same software ("MediaWiki") as Wikipedia and considered adopting and improving its content.[38] However, definitively settling upon Citizendium's license was not a quick or easy process. One concern among some Citizendium contributors was that if they were to use the GFDL license, and therefore able to use (and improve upon) Wikipedia content, Wikipedia could import the improved Citizendium content back into itself. This was unacceptable to those who wished to distinguish themselves and the superiority of their approach. Therefore, as some Wikipedia content had already been adopted, Citizendium experimented with the possibility of "unforking" their borrowed content—rewriting it from scratch. However, in December 2007, Citizendium chose a Creative Commons license, which, after an orthogonal effort to make it and the GFDL compatible, means "Wikipedia and the Citizendium will be able to exchange content easily."[39] In any case, forking of Wikipedia is an acknowledged possibility and consequent of its openness, as Wales noted early on with Nupedia: "One important thing to note is that if advocates of some viewpoint wish to claim that we are biased, and we are unable to come to a consensus accommodation of some kind, they will be free to use our content as a foundation, and to build their own encyclopedia with various articles added or removed. This is the sense in which open source is about free speech, rather than free beer."[40]

Discussing Openness

Openness entails a handful of constituent values that are not always easily reconcilable in problematic situations. In the following sections I review four cases that challenged the community on questions of openness: whether anyone can really edit, the legitimacy of office actions, the effects of bureaucratization, and the WikiChix enclave.

Can Anyone Really Edit?

As noted, the English Wikipedia declares itself as "the free encyclopedia that anyone can edit."[41] Presently, this includes the unregistered/anonymous (i.e., those who don't have an account or don't log into it before editing). Despite the common retort that Wikipedia is "not an anarchy," among other things,[42] the feature of openness and anonymous editing continues to be a valued part of Wikipedia's identity: even those who always log in might still support allowing others to edit without logging in.

Before discussing anonymity, blocking, and openness, some background information is in order. Every edit to Wikipedia is captured and can be reviewed on the article's history page. Wikipedia contributors may choose to register an account with a name/identity of their choosing: it might be personally identifiable, or a pseudonym. Editors who have not logged in to such an account are often referred to as "anonymous."[43] In the history log, the edit of an anonymous user is attributed to an IP address, the number associated with a user's computer by their Internet service provider. The reason the term *anonymous* is not strictly correct is that there have been cases in which these numbers have been traced back to a particular computer. For example, the offices of U.S. Congressional representatives were identified as being responsible for removing true but embarrassing information on Wikipedia.[44] In fact, individuals who wish to protect their privacy would be better off creating a pseudonym with which to edit Wikipedia content. Then, only the few "checkuser" Wikipedians would be able to determine the IP address of the originating computer.[45]

Unfortunately, Wikipedia is continually vandalized. However, there are various automated tools ("bots") and groups of users (e.g., the "RC [recent changes] Patrol") that roll back or "revert" articles to their previous state. If a page is particularly contentious, it can be *protected* for a period to prohibit all nonadministrator changes. When it becomes clear that a specific user is persistently abusive, administrators may *block* his or her editing for a specified period, or in serious cases they might institute a lifetime *ban.*[46] However, a blocked user might create another account or edit anonymously. Consequently, administrators have the ability to block users based on their IP addresses. While some blocks might be mistaken or questionable, it is difficult to conceive of Wikipedia working without such a feature as it would soon be overwhelmed with junk. As the "Wikipedia Is Not an Experiment in Anarchy" article states:

> Wikipedia is free and open, but restricts both freedom and openness where they interfere with creating an encyclopedia. Accordingly, Wikipedia is not a forum for unregulated free speech. The fact that Wikipedia is an open, self-governing project does not mean that any part of its purpose is to explore the viability of anarchistic communities. Our purpose is to build an encyclopedia, not to test the limits of anarchism.[47]

However, since one's IP address can change, or many users may share an IP address, this approach sometimes blocks the innocent; a balance must be struck between the values of openness and quality. Specifically, to what extent do technical features such as blocking vandals or requiring registration promote or constrain community values such as openness?

Consider an infamous case of 2005 in which the biographical article of John Seigenthaler, an administrative assistant to Robert Kennedy, contained the unfounded claim that he was implicated in the assassinations of the Kennedy brothers. Much to the embarrassment of many Wikipedians, Seigenthaler objected in a widely discussed editorial opinion in *USA Today.*[48] After the identity of the "anonymous" contributor was revealed as the author of a "prank gone wrong," the press reported that Seigenthaler was not holding a grudge, or supporting a regulatory crackdown on the

Internet, but he did fear "Wikipedia is inviting it by its allowing irresponsible vandals to write anything they want about anybody."[49]

In a message to one of the Wikipedia lists, Jimmy Wales objected to this as a mischaracterization of Wikipedia, and its openness. Wales argued that to equate openness with defamation is like equating a restaurant's steak knives with stabbings. To force everyone in the restaurant to be isolated in steel cages because of the possibility of a stabbing would violate the values of "human kindness, benevolence, and a positive sense of community" and, consequently, "I do not accept the spin that Wikipedia 'allows anyone to write anything' just because we do not metaphysically prevent it by putting authors in cages."[50] The question here seems to be to what extent does the phrase "allow anyone to edit" include the possibility of "allowing pranksters to defame"? Seigenthaler's position seems to be that the freedom to edit implies abuse, whereas Wales seems to be arguing that such a conclusion is misleading with respect to the community's intention and the balance of consequences. While vandalism is actively resisted by Wikipedians, implementing technical or social structures that would make vandalism *impossible* would conflict with other community values. How might Wikipedia decrease the possibility of vandalism without unduly affecting other values such as openness?

In December 2005, shortly after the Seigenthaler incident, a new mechanism was announced: page "semi-protection." This prevents unregistered editors, or those registered within the last four days, from editing the protected page. Previously, any user, anonymous or logged in, would be prohibited from editing a "fully" protected page. Now, while semi-protected pages can't be edited by anonymous and new users, established users can edit them.[51] Hence, moving pages from full protection to semi-protection makes the pages more accessible to Wikipedians.

Much to the chagrin of the community, this proposal gained major attention with the publication of a *New York Times* article entitled "Growing Wikipedia Revises Its 'Anyone Can Edit' Policy."[52] This led Wales to comment that:

> Not every case of allowing more people to edit would count as "more open." For example, if we had a rule that "Only Jimbo is allowed to edit this article" then this would be a lot LESS open than "no one is allowed to edit this article." Openness refers not only to the number of people who can edit, but a holistic assessment of the entire process. I like processes that cut out mindless troll vandalism while allowing people of diverse opinions to still edit. Those are much better than full locking.[53]

On June 21, 2005, the *New York Times* corrected its original article by noting that some form of protection had always existed on Wikipedia, and the online version's headline now reads "Wikipedia refines" its policy, rather than "revises" it.[54] And semi-protection was soon complemented with the ability to "soft block" IP numbers, meaning one could block anonymous users from troublesome IP addresses, "but allow editing by registered users when logged in."[55] (Remember, one *protects* specific pages, but *blocks* users.) During the lengthy discussion about the feature, some expressed a concern that it was contrary to the value of openness:

> Personally, I think the new blocking policy . . . will do more harm than good. The proposal would indubitably mean the blocking (using this logged-in only registration) of most AOL IPs, Netscape IPs, school districts, public-use computers, and major corporations. By only allowing logged-in users on these IPs (since it is inevitable that all of them would either be blocked indefinitely or blocked consistently), in my opinion, is against the spirit of the Wiki—we're here to allow *anyone* to edit, not just those who want to create accounts.[56]

Others countered with a pragmatic argument. On the face of it, it might appear there are more restrictions as there is a new feature in the software, yet it would further the goal of greater access in practice:

> I really can't figure out what you're arguing here, though. Because right now, when an AOL IP is blocked, you can't edit using it regardless of whether or not you register. As I understand it, the proposal is to allow logged in users to edit when they otherwise wouldn't. Sure, this might lead to admins being more liberal with IP blocks, but it doesn't require it—whether or not admins are more liberal with IP blocks is a separate issue, and we could pass policies to ensure that this doesn't happen.[57]

In July 2006 acceptance of the soft-blocking proposal was characterized as an "avalanche" of support and I have seen little evidence that it or semi-protection has negatively affected users.[58]

How does this story of anonymous users, vandals, and blocking engage the idea of Wikipedia as an open content community? There are four issues worth explicitly identifying so as to answer this question.

First, what is the scope of "anyone"? Does "anyone" include persistent vandals with no goal other than disturbing Wikipedia? The community has comfortably concluded that it does not—though it does continue to be quite forgiving by preferring suspension and a process of escalation before outright banning occurs. Does "anyone" include anonymous editors? Historically it has, and continues to do so except in cases of suspected abuse.

Second, how to balance values? As social software researcher Clay Shirky points out, a group needs the *right amount* of freedom; new collaborative tools, such as the wiki, can enable, but not guarantee, a balance that is neither overly managed nor chaotic.[59] And openness is not the only value of Wikipedia, it is not even the primary one. The ultimate goal of Wikipedia is to produce a high-quality encyclopedia. Many believe openness furthers the ultimate goal of producing quality content, but a quality encyclopedia should not be sacrificed in the face of a detrimental openness. Fortunately, the values of openness, quality, and kindness are often seemingly sympathetic to each other. Yet, as seen, there are cases in which they are in tension and can be addressed through additional technical intervention.[60] Sanger, with the Citizendium project, for example, has chosen a different balance by requiring that all contributors use their real-world identities.

Third, does *possibly* imply *essentially*? In an expansion of an argument by Langdon Winner wherein certain technologies (e.g., nuclear) can be inherently political (i.e., inherent to certain social and political relationships),[61] some critics maintain that because certain things are *possible* on Wikipedia they are *essential* to Wikipedia. Whereas Winner argues that the dominant uses of technology are determinative, Wikipedia critics go further and argue that even a possibility is determining, or to put it another way, "because Wikipedia permits foo, it is foo-ish." Others respond that marginal cases do not define the whole and should not be catered to if they conflict with more central values. To this end, Wales was quoted in the *New York Times* article as saying: "Protection is a tool for quality control, but it hardly defines Wikipedia. What does define Wikipedia is the volunteer community and the open participation."[62]

Fourth, do technological constraints always imply movement away from openness? The ability to block anonymous users associated with an abusive IP address was a new feature. Yet innocent anonymous users would have been blocked before, as would have those users logged in at that IP address. With the new feature the latter group has access it did not before. In this case we see the relevance of historical context (existing practice) and practical effect on the meaning of a technical feature.[63]

Although some might argue any effort to block even problematic users is a step away from openness, a chaotic culture of undisciplined vandals would equally disenfranchise those who wish to make a positive contribution to a viable encyclopedic project. The community must undertake a balancing act, one that is difficult, occasionally settled, and then disrupted again. For instance, in 2009, because of continuing concerns about inappropriate edits to biography articles, Wales proposed that the "Flagged Revisions" feature be enabled on the English Wikipedia.[64] Flagged revisions are a long-discussed mechanism by which an approved—rather than the latest—version of a page would be seen by "the public" (i.e., those not logged in). This could be used to create a higher-quality view of a wiki, or present a stable and inoffensive view of a contentious article (i.e., "Flagged Protection"[65]). This proposal might also be thought contrary to the wiki ethos of one's edits being seen immediately upon hitting the "save" button. Indeed, in Wales's "Statement of Principles" he wrote "'You can edit this page right now' is a core guiding check on everything that we do. We must respect this principle as sacred."[66] Of course, sometimes the balance must be shifted and one could still *edit* a page "right now" even if it is not immediately *seen* by the public. Furthermore, while this solution might look like a constraint, a closing down of a wiki, it could very well provide greater access than if a page is simply protected.

Interfacing with the Outside World

On Wikipedia one is expected to discuss the editing of an article with fellow contributors, at a minimum by including a summary when saving an edit. Arguments are made in the open with reference to verifiable sources and community policy. However, for those with a proprietary interest, this process of reasoned discussion can be circumvented via a call or letter to the "Wikipedia office," that is, by formally contacting the Wikimedia Foundation. And, sometimes, rightfully so. What obligation did Seigenthaler, someone completely unfamiliar with wikis, have to remove the libelous claim from Wikipedia that he was implicated in the assassination of the Kennedys?[67] None. As Wales wrote, "The problem we are seeing, again and again, is this attitude that some poor victim of a biased rant in Wikipedia ought to not get pissed and take us up on our offer of 'anyone can edit' but should rather immerse themselves in our arcane internal culture until they understand the right way to get things done."[68]

However, unfortunately, the office mechanism can be abused by those pushing a nonencyclopedic point of view (POV), such as promoting (or censoring negative views of) a commercial product. If such people can't win their arguments on the merits of notability, verifiability, and neutrality within the community, having their lawyer call the office might prompt an office intervention—such as deleting objectionable material, which would then be labeled with the WP:Office template.

Something like WP:Office, where discussion occurs off-wiki, is an unfortunate though (probably) necessary mechanism for avoiding legal problems. Yet, in an ironic twist, office actions soon became a red flag to those who dislike this intervention or otherwise like to make trouble for Wikipedia (e.g., copying sensitive or contentious materials off Wikipedia to continue a controversy). Whereas office actions were intended to quickly and quietly remove a potential liability, they became a flashpoint. This led to the unfortunate case in which office actions were taken without being labeled as such, so as not to draw attention, and a prominent user had his administrator status revoked and was blocked indefinitely because he had reverted an unlabeled office action. (In the end, his response was an exemplar of Wikipedia tact and his position was soon restored.[69]) Perhaps because of the attention caused by office actions, a *suppression* (revision-hiding) feature was introduced that "expunges information from any form of usual access even by administrators," including "username, revision content, and/or edit summary in order to remove defamatory material, to protect privacy, and sometimes to remove serious copyright violations."[70] Access to this feature is limited to the few users given "oversight" permission. This is even a greater step from openness and is evidence that legal threats had clearly become a top priority for the Wikimedia Foundation. Indeed, in 2006 the Wikimedia Foundation hired one person to be both "general counsel and interim executive director."[71] Similar concerns over the incompatibility of copyright and liability regimes with open content production have led Larry Sanger to argue that works "developed in a strongly collaborative way" merit special protection under a law that is sensitive to the novel way in which they are produced.[72]

Wikipedia's suppression of information became a mainstream story in June 2009. David Rohde, a reporter for the *New York Times*, had been kidnapped by the Taliban and held in Afghanistan for seven months until he managed to escape. When he did escape, the *Times* reported that it had suppressed the story during that time, and asked others to do the same, so as not to draw attention and further endanger Rohde. This included a request to the Foundation to help keep this information off of Wikipedia as an anonymous contributor was attempting to add it to Rohde's biographical article. Knowing that his actions drew attention, Jimmy Wales worked with others to prevent its reappearance. Removals were justified based on the lack of sufficiently reputable sources for the claim—since the media was suppressing the story—and the page was subsequently protected (i.e., locked from further edits). Wales said they had no idea who this contributor was and so "there was no way to reach out quietly and say 'Dude, stop and think about this.'"[73] On June 20, the news broke and Wales removed the protection. In less than thirty minutes, the contributor, still not appreciating why the information was suppressed, added it again with references to stories of Rohde's escape with an edit summary that said: "Is this enough proof you fucking retards? I was right. You were WRONG.:P"[74] Within the community, but mostly from without, there was much discussion about Wikipedia's adherence to its principles of openness and veracity, and its responsibility to the safety of others since it is not only subject to influence from the outside world, but can also affect the world it documents.

Bureaucratization

The community's own internal development also has implications concerning its openness. Sociologist Max Weber observed that leadership often shifts from a charismatic leader to a more bureaucratic form of governance as a community matures.[75] Clay Shirky, a contemporary scholar of "social software," makes a more irreverent observation: "Process is an embedded reaction to prior stupidity," meaning "an organization slowly forms around avoiding the dumbest behaviors of its mediocre employees, resulting in layers of gunk that keep its best employees from doing interesting work."[76] The Wikipedia essay "Practical Process" clearly defines the role of process (i.e., to implement policy, to provide consistency, to reduce redundancy in similar situations, and to further learning and decision making); but it also identifies how it emerges, and hardens, how to recognize bad process, and warns, "You can't legislate against misunderstanding or malice."[77] (An earlier version of the essay was more concise: "You can't legislate clue."[78])

Wikis do not add unnecessary process in and of themselves: they are simple, accessible, flexible, quick, and cumulative. Furthermore, community process need not be overly specified in fear of a mistake since content changes are easily reverted. (Trusted leadership also plays an important role here.) However, an unforeseen implication of the wiki's ability to facilitate content creation is that policies are but another type of content. So, in the end, Wikipedia is no exception: despite the norm of "Avoid Instruction Creep," the "ratio of policy citations to talk edits" is increasing.[79] For example, Andrew Lih, author and Wikipedia administrator, prompted a discussion over what he saw as an overly officious statement about the "speedy deletion" of a page he found useful. The deletion notice warned that the article, "is a very short article providing little or no context (CSD A1), contains no content whatsoever (CSD A3), consists only of links elsewhere (CSD A3) or a rephrasing of the title (CSD A3)." Lih responded:

> It's incredible to me that the community in Wikipedia has come to this, that articles so obviously "keep" just a year ago, are being challenged and locked out. . . . It's as if there is a Soup Nazi culture now in Wikipedia. There are throngs of deletion happy users, like grumpy old gatekeepers, tossing out customers and articles if they don't comply to some new prickly hard-nosed standard. It's like I'm in some netherworld from the movie Brazil, being asked for my Form 27B(stroke)6.[80]

Some degree of policy is necessary in any community, and bureaucratization is a common—many would say unavoidable—feature of organizational development. As one study of Wikipedia policy concludes, "the 'policyless' ideal that wikis [supposedly] represent is a pipe dream."[81] Yet, even in the face of a proliferation of process, the open content community values of transparency and integrity are largely preserved. However, should the accretion of policy become too heavy, integrity can be compromised by frustration and "Wikilawyering" (i.e., employing overly technical or legalistic arguments that focus on the letter of policy rather than its spirit).[82] For example, the policy boom has prompted one Wikipedian to declare that he had "Kicked the Process Habit": "So as of today, I'm just going to go ahead and edit. Lord knows the rules are making me nervous and depressed. So I'll follow all the stuff I can remember, and not try too hard to learn the other stuff."[83]

Enclaves and Gender

One should not be surprised that a source of contention in open content communities is when a subset of community members creates a closed space. The conditions that prompt such proposals and the rhetoric marshaled to support or attack them give insight into a community's attempts to understand and implement openness.

A common feature of online communities operating under an ethos of open and egalitarian values is frustration with the coexistence of group decision making and seemingly contrary forms of autocratic authority. (I use *authority* to describe the right to exercise *power*, which is in turn the capacity to influence;[84] by *autocratic* I mean actions that do not derive their authority from group decision-making processes.) Evidence of this phenomenon includes the alleged "secret cabal" of Usenet in the 1980s, private "sysop"-only email lists or IRC channels, and the "benevolent dictators" of communities including Python, Linux, and Wikipedia.[85] For example, consider the following comment on a Wikipedia email list: "There are many private, semi-private and secret lists in which wikimedians make decisions with each other without ever telling anyone or explaining. Openness has gone overboard a very long time ago. Most things you read on the public lists have been discussed privately long before an outsider found out about them."[86]

However, in 2006, a different sort of closed group was set up outside of the larger Wikipedia community, what might be called an enclave or minority-specific space.[87] For while cabal formation is a seemingly inevitable structural result of group decision making, and legal threats are an inescapable reality of living in a litigious society, enclaves are purposely chosen by a subset of the community in seeming contradiction with the values of openness and equality. This was aptly demonstrated in the Wikipedia community by the announcement of a "WikiChix" list for female-only discussion, which began when:

> Offlist chat about the recent discussions on systemic gender bias in Wikipedia made it clear that a number of women were not comfortable contributing to the conversation there. This inspired the creation of WikiChix in November 2006. WikiChix is a wiki and mailing list for female wiki editors to discuss issues of gender bias in wikis, to promote wikis to potential female editors, and for general discussion of wikis in a friendly female-only environment.[88]

Formally excluding anyone from the larger community prompts questions of fairness and discrimination. Some members reacted by arguing of a slippery slope toward absurdity, such as a need for "a mailing list for homosexual African-Americans from planets other than earth."[89] In a similar spirit, another Wikipedian asked about the need for a "British-only or atheist-only" list but also acknowledged the specific motivations for the creation of WikiChix: "the list was organised to avoid a specific problem—women feeling uncomfortable posting to this male-dominated list where explicitly sexist statements (even if they weren't meant seriously) are left unchallenged by a large number of people."[90]

Earlier, I noted that encyclopedias and FOSS were both inspirations for Wikipedia. However, both domains have been historically male dominated and this too is part of Wikipedia's inheritance. The well-known gender imbalance in computer-related fields is further exacerbated in the FOSS community, with females making up only about 1 percent of participants.[91] In fact, the very notion of equality may inhibit constructive action toward mitigating bias. After interviewing male and female students about computer usage and its larger culture (e.g., reading computer magazines), Fiona Wilson argues that women who might otherwise object to informal bias might simply accept the presumption of equality or not want to challenge it so as to avoid being singled out.[92] The model of female "chix" projects (e.g., LinuxChix, Ubuntu Women, Debian Women, KDE Women, WikiChix) appears to be a positive counterforce to this tendency.

Another response employed by those concerned with such spaces is not to object to the exclusion, but to the division of the larger community. Shouldn't the community ensure the common space is accessible rather than spinning off groups? For example, "A better solution would be to kick any of the men that behave like that, not to assume that 'all men are chauvinist pigs.'"[93] Unfortunately, the history of reference-work production is rife with examples of male chauvinism. Robert Cawdrey's 1604 "Table Alphabetical" can't help but be read today as patronizing given it was "gathered for the benefit & helpe of Ladies, Gentlewomen, or any other unskilfull persons."[94] In early encyclopedias, women often merited only a short mention as the lesser half of man. A notable exception, the article on midwifery in the first edition of *Britannica* and its illustrations of the female pelvis and a fetus prompted a public scandal. King George III ordered the forty-page article on midwifery destroyed, pages and plates.[95]

Furthermore, few women prominently appear in the historical record of reference works. The few exceptions are in the domain of librarians and documentalists, such as Suzanne Briet and her peaceful reading room. Unfortunately, even Melvil Dewey's advocacy for women in the library profession is marred by alleged discrimination and personal scandal.[96] This juxtaposition of limited advances in the context of continuing bias is also a theme in Gillian Thomas's history of women contributors to the *Britannica*, the only book I've found so far to address this issue directly.[97] Today, at least, women are visible in everyday tasks and positions of authority at Wikipedia, though they continue to be a minority (roughly 10 percent according to surveys[98]) and concerns of "systematic gender bias" persist.

Of course, given the value accorded to free speech, the community would have a difficult time restricting the speech of "men who behave like that." How would such a determination be made? One of the few standards available for the discrimination of speech in online communities is that of *trolling*, a term describing contentious speech, probably not even genuinely held, that is expressed for the sole purpose of inflaming discussion. But how would one distinguish between misogyny and trolling?[99] (Or, how does one distinguish between genuine racism and provocation? Offensive statements used to antagonize others in a heated moment need not be believed.) An irony is that falsely held misogynistic statements espoused for the purposes of trolling might be censured or censored, but a genuine misogynist could claim that any formal censure is a form of "thought crime," which is generally anathema under free-speech principles.

This type of discussion that traverses the difficult questions of freedom and equality often prompts extensive debate. Although discussions about these values sometimes create a shared "productive ethical orientation" within the community,[100] they can also be alienating and seemingly endless. This is why such topics are so suitable to trolling in the first place, and for which community leaders often step in, as Wales notes:

> The point is, if the broad philosophical question is "Do we ban people for merely holding unpleasant or unpopular beliefs?" then the answer is "no, we never have, and there seems to be very little support for doing so." If the point is "Does asserting unpleasant or unpopular beliefs automatically get you a free pass to be any sort of jerk you like, because we are planning to bend over backwards to make sure we don't ever ever ever discriminate against Nazis?" then the answer is, "no, being a disruptive troll is still being a disruptive troll."[101]

Not surprisingly, it did not take long for the WikiChix proposal to be challenged; a longtime male contributor, and self-described "overly combative" anarchist,[102] tried to subscribe to the list and was rejected. (I suppose this action was a violation of the norm "Do Not Disrupt Wikipedia to Illustrate a Point,"[103] which brings some measure of sanity to difficult issues.)

The final, parliamentary, objection to the WikiChix proposal was that this exclusive list was being hosted by the Wikimedia Foundation. The other free software-related women fora, while focused on being "women-friendly," are more or less open and affiliated with the larger community.

LinuxChix "is intended to be an inclusive group where everyone is and feels welcome. . . . LinuxChix is intended to be primarily for women. The name is an accurate reflection of that fact. Men are welcome because we do not want this group to be exclusive."[104] Debian Women states: "We're not segregated. Debian Women is a subgroup of Debian that allows anyone to join and help."[105] On UbuntuWomen, "Membership is open to all."[106] The KDE Women Web site is run by women so "you have to be a woman,"[107] but in addition to the six listed female members, there are also five male "supporters" and men are present on the IRC channel and mailing list. The gender exclusivity of WikiChix is atypical and it is not clear to what extent this decision was purposeful and what the consequences might be relative to the other female-friendly fora.

In the end, the WikiChix list was moved from being hosted by Wikimedia, which might carry the presumption of endorsing exclusive discrimination, to a non-Wikimedia host. In response to this, a Wikipedian responded: "Excellent. I still think it's a bad idea, but if it's not being supported in any way by Wikimedia Foundation there's no need to complain about it here any more."[108] As is often the case on difficult issues, the conclusion to this argument was facilitated as much by exhaustion as by reason. Endless argument about whether bias exists, rather than partaking in constructive dialogue on how to counter it, is a reason such spaces are often created. By severing any support and official affiliation with the Wikimedia Foundation the WikiChix list became moot to the larger community.

While this particular case was resolved by simply moving the list, it still is illustrative of a challenge to openness. As Freeman notes, informal—though no less exclusionary—boundaries may persist despite the absence of formal exclusions.[109] Therefore "formal" enclaves can be a productive response to the "tyranny" of informal structures and biases of a larger community. Legal scholar Cass Sunstein recommends that in such circumstances, "it can be indispensable to allow spaces in which members of minority groups, or politically weak groups, can discuss issues on their own. Such spaces are crucial to democracy itself."[110] Critical theorist Nancy Fraser writes of a similar notion, "subaltern counterpublics," wherein subordinated social groups can formulate and discuss interpretations of their identities, interests, and needs counter to the mainstream discourse.[111]

Yet, WikiChix's exclusion of males, rather than being open and pro-female, is obviously problematic with respect to transparency and nondiscrimination. Also, Sunstein recognizes that enclaves can further group polarization and marginalization and recommends that enclave members be brought back into contact with the larger community; otherwise, self-insulation can yield extremism.[112] In following this issue I haven't perceived a decrease in female presence after the provision of a female-friendly space. A counter to the hypothesis that women are abandoning the common space is the hypothesis that having a more supportive space to fall back on will encourage comfort in speaking in common spaces. Yet these other female-specific spaces are also open, whereas WikiChix is gender exclusive. In the end, time will tell, and I expect that because all constituencies still possess a common object (Wikipedia), marginalization and extremism will be minimal.

Conclusion

Wikipedia is an example of an *open content community*. Such a conceptualization entails the core value of providing open content, and the implication of forking. However, it can be difficult to balance the associated values of transparency, integrity, and nondiscrimination, as well as other concerns such as free speech and the safety of people and the project itself. Furthermore, boundaries are a fundamental feature of any community, even for those that aspire to openness because it is rarely a simple binary of open or closed. Even a theoretically perfect openness can lead to behavior and informal structures that are less than inclusive. As Clay Shirky writes, "successful open systems create the very conditions that require and threaten openness. Systems that handle this pressure effectively continue (Slashdot comments). Systems that can't or don't find ways to balance openness and closedness—to become semiprotected—fail (Usenet)."[113]

This is the sort of insight not present in H. G. Wells's predictions of a Modern Utopia, Open Conspiracy, and World Brain, but emerges when one spends time in the Wikipedia community. Ultimately, an important descriptive feature of an open content community is a lot of discussion about its values and how to balance them. By this measure, Wikipedia certainly qualifies.

5 The Challenges of Consensus

Consensus: Any group in agreement about something whose opinion is the same as yours; antonym of cabal [i.e., those who disagree with you].
—WikiSpeak

H. G. Wells thought the "World Encyclopedia" should be more than an information repository, it should also be an institution of "adjustment and adjudication; a clearinghouse of misunderstandings."[1] Wikipedia certainly has its share of misunderstandings, some imported from the conflicted world it documents and some unique to its own undertaking. An example of a contagious real-world conflict is the "Creation-Evolution Controversy,"[2] discussed in chapter 3. Also, political and ethnic differences are often mirrored at Wikipedia, prompting the formation of a "Working Group on Ethnic and Cultural Edit Wars."[3] There are also plenty of local "misunderstandings," such as whether every episode of *Buffy the Vampire Slayer* deserves its own article. I raised this dispute earlier to illustrate two opposing philosophies at Wikipedia: inclusionism and deletionism. This issue, and the proliferation of articles, gave rise to an even more trivial—though no less bellicose—debate: if every television show episode has its own article, how should these articles be named so as not to conflict with other articles? This discussion reveals possible misunderstandings about consensus, and the difficulties of this decision-making practice in an open community.

In this chapter, I identify the difficulties of consensus decision making, and its meaning and practice for collaboration at the English Wikipedia. I consider this relative to insights from literature about consensus in other communities, including Quakers and the collaborators who built the Internet and Web using "rough consensus and running code."

The Case of Disambiguation

In the history of the encyclopedia much has been made of the attempts to organize knowledge and how that dream was eventually superseded by simple alphabetical order.[4] Wikipedia has continued this trend by avoiding any formal organizational scheme and letting people simply name articles as seemed fitting; articles are then accessed via other pages, including user-created categories and search engines.[5] But a problem soon emerged: what happens when article titles conflict? A page's title must be unique because it is also part of the Web address of the page. (Computer scientists call this a "collision" and it became increasingly common as the number of Wikipedia articles increased.) For example, what should the "Buffy" article contain? Should it be about the fictional character, Buffy the Vampire Slayer, or the related film, TV show, comics, or novels—and which season or episode, issue, or book? Plus, there have been a few notable real-life Buffys. How should Wikipedia distinguish between them all?

In the case of a collision Wikipedia will likely offer the reader a "disambiguation" link at the top of an article or a whole page with a list of links to more specific articles, or both. "Disambiguation in Wikipedia is the process of resolving conflicts in article titles that occur when a single title could be associated with more than one article. In other words, disambiguations are paths leading to different articles which could, in principle, have the same title."[6] The "Buffy" article is in fact a disambiguation page that includes links to articles about the *Buffy the Vampire Slayer* film and television series, unrelated musical albums, biographies, and an astronomical object outside the orbit of Neptune.[7] Furthermore, dozens of naming conventions have emerged that specify how to disambiguate collisions by qualifying the name with a parenthetical suffix, such as "Buffy the Vampire Slayer (film)."

However, must these disambiguators be applied in every case for consistency's sake, or just those for which there is already a preexisting article? Answering this question, and thousands of others like it, is an integral part of Wikipedia collaboration. As the "Consensus" policy states, consensus is "how editors work with others," it is "Wikipedia's fundamental model for editorial decision-making." Wikipedians are supposed to discuss and reason together, making use of verifiable sources and assuming good faith. "Policies and guidelines document communal consensus rather than creating it."[8] However, in this case of disambiguation an interesting problem arose: there was disagreement as to whether there was consensus.

So then, what is consensus and how do you know you've reached it? This question eventually reached the Arbitration Committee: "a panel of experienced users that exists to impose binding solutions to Wikipedia disputes that neither communal discussion, administrators, nor mediation have been able to resolve."[9] Jimmy Wales created the committee in 2004 because of the growth of the community and the corresponding number of disputes he could no longer personally attend to. Wales appoints members annually "based on the results of advisory elections" to arbitrate specific conflicts; however, the "ArbCom" seems to be evolving toward what might be thought of as Wikipedia's high court in definitively interpreting—and some say making—Wikipedia policy.[10]

Presently, an ArbCom decision is documented on a wiki page in which disputants make statements and marshal evidence for their case.[11] These are followed by the ArbCom members' "preliminary decision" (e.g., to take the case and issue temporary injunctions) and conclude with a final decision that enumerates important principles of Wikipedia policy, findings of facts, remedies, and enforcement actions. Wikipedian Yaksha introduced the disambiguation case as follows:

> This dispute is regarding whether articles for TV episodes which do not need to be disambiguated should have disambiguation. For example, Never Kill a Boy on the First Date (Buffy episode) has the disambiguation "(Buffy episode)," even though this disambiguation is not required. I believe we did reach consensus to follow the existing guideline of "disambiguate only when necessary." The straw poll resulted in a supermajority (80%) support for "disambiguate only when necessary." The discussion that followed supported this consensus. A detailed summary of the discussion, as well as four Request Move proposals all support the existence of this consensus. Given this, I (and others) begun to move articles which were inappropriately named.[12]

However, not everyone agreed with the application of the use-as-needed policy and attempted to reverse the moving (or renaming) of articles. Wikipedian Elonka was of a minority that felt specific "WikiProjects" (i.e., pages and editors focused on advancing specific topics, such as a television series) should be able to use disambiguating suffixes consistently across their topic, whether needed or not. Putting aside "unethical tactics" that had been employed in the dispute, this minority that favored consistent-suffixes felt that "WikiProjects can set reasonable guidelines of their own."

"As television episode articles have been added to Wikipedia, most series followed the [use as needed] system, but many others chose to use a 'consistent suffix' system."

Furthermore, the very process of deciding whether specific WikiProjects (e.g., television) could consistently use suffixes was troubled. Elonka noted that there was a poll, but its wording was confusing and contested "with multiple editors rapidly changing the wording and structure of the poll while it was in process." Furthermore, "calls for a cleanly-run poll were belittled as 'stalling', 'immature delay tactics', and the 'whining' of 'sore losers' engaging in 'borderline trolling' who should just, 'Give the fuck up, you lost.'"[13]

Had there been consensus on the naming of television episodes? Before returning to the details of this case, it's best to first review the meaning of consensus and its seminal role in the development of the Internet.

"Rough" Consensus

The Wiktionary definitions for consensus speak of "general agreement," "without active opposition to the proposed course of action." A more scholarly source gives a similar definition: consensus is overwhelming agreement "which does not mean unanimity."[14] The encyclopedic article "Consensus Decision-Making" lists requirements of consensus that, if achieved, can also be considered benefits: inclusive ("as many stakeholders as possible"), participatory ("actively solicit the input and participation of all"), cooperative ("reach the best possible decision for the group and all of its members"), egalitarian ("equal input" with the "opportunity to table, amend and veto or 'block' proposals"), and solution-oriented ("emphasize common agreement over differences").[15] This is not unlike the meetings of one of the better-known practitioners of consensus: the Quakers. Michael Sheeran, a Jesuit scholar, writes of the history and practice of Quaker consensus in *Beyond Majority Rules: Voteless Decisions in the Religious Society of Friends*. In his study Sheeran notes nine features of Quaker meetings and decision making.[16] One of those characteristics, central to the Quaker spiritual experience, has no analog in Wikipedia: silent periods at the start of meetings and when conflict arises. The characteristic of "small meetings" sometimes holds in the Wikipedia context for issues local to an article or project, but not at the larger scale.[17] The remaining seven characteristics roughly parallel Wikipedia norms: unanimity and a lack of voting (e.g., "voting is evil"); pausing when agreement cannot be reached; participation by all those with ideas on the subject; listening with an open mind; facilitators, but no "leaders"; egalitarianism; and a factual rather than emotional focus.

Therefore, consensus certainly seems like an appropriate means for decision making in a community with egalitarian values and a culture of good faith. Furthermore, this form of decision making has been central to online collaboration since the Internet's start. Yet, while consensus might seem simple enough in theory, it is rarely so in practice, as is evidenced by the 1,176 pages of *The Consensus Building Handbook*.[18] The history and challenges of online consensus, particularly this question of who decides when one has it, can be seen in the development of technical standards at the Internet Engineering Task Force (IETF) and World Wide Web Consortium (W3C).

Both of these institutions host technical working groups that develop standard specifications for Internet protocols (e.g., TCP/IP) and Web formats (e.g., HTML). They work primarily over mailing lists, though teleconferences and occasional face-to-face meetings are common. The IETF is one of the oldest existing collaborative institutions online—it can be said to have built the Internet. And the W3C, responsible for many Web technologies, might be thought of as an institutional fork resulting from, in part, frustration over the slow pace of work at the IETF. Much of this frustration was a result of trying to come to consensus over technical philosophical differences. One of the most contentious issues had to do with naming/identifying things on the Web—further evidence that naming things is not as easy as one might initially think. In this case the disagreement was about whether the string of characters one types into the address bar of a browser should be thought of as a stable identifier for that Web resource (i.e., URI) or just a locator (i.e., URL).[19] (This distinction is confusing, might seem trivial to most, and has become less of an issue, but it was of great concern to those involved at the time.) And while the W3C, unlike the IETF, has a paying membership that helps support a full-time staff (to hopefully speed the work along) and has Tim Berners-Lee as director to lend coherence and direction to Web architecture, consensus decision making at the working group level was retained.[20] The W3C process document states: "Consensus is a core value of W3C. To promote consensus, the W3C process requires Chairs to ensure that groups consider all legitimate views and objections, and endeavor to resolve them, whether these views and objections are expressed by the active participants of the group or by others (e.g., another W3C group, a group in another organization, or the general public)."[21] Ironically, as the W3C matured, it too would be characterized as overly slow because of growing bureaucracy and the difficulty of achieving consensus in large and interdependent groups. Furthermore, Berners-Lee's leadership role, which was intended to mitigate these problems and lend architectural coherence to the emerging standards, was occasionally challenged as not in keeping with the consensus practice of the working groups.[22] In turn, Jon Bosak, a "father" of XML, a data-markup and exchange format—and one of W3C's most prominent successes—created an institutional fork for subsequent XML work. The Organization for the Advancement of Structured Information Standards (OASIS) jettisoned the ideas of appointing a director and operating by consensus in favor of the parliamentary, and more clockwork-like, Robert's Rules of Order.

In any case, it was at the IETF in 1992 that computer scientist David Clark characterized IETF collaboration: "We reject: kings, presidents and voting. We believe in: rough consensus and running code."[23] This "IETF Credo" would become one of the foundational aphorisms of collaborative culture on the Internet. Furthermore, this simple statement reflects the egalitarianism—and meritocracy—described in previous chapters of this book. It also hints at a source of skepticism of some Free and Open Source Software (FOSS) developers toward Wikipedia since encyclopedic articles cannot be compiled and "run." Returning to the question of whether 80-percent support for a Wikipedia policy constituted consensus, in the IETF Credo we also see the intriguing notion of "rough consensus" as documented in the IETF's "Working Group Guidelines and Procedures":

> IETF consensus does not require that all participants agree although this is, of course, preferred. In general, the dominant view of the working group shall prevail. (However, it must be noted that "dominance" is not to be determined on the basis of volume or persistence, but rather a more general sense of agreement.) Consensus can be determined by a show of hands, humming, or any other means on which the WG agrees (by rough consensus, of course). Note that 51% of the working group does not qualify as "rough consensus" and 99% is better than rough. It is up to the Chair to determine if rough consensus has been reached.[24]

Yet even 70 percent of a group, for example, as determined by a chair seems like a far cry from "general agreement without opposition" as described earlier. Indeed, to understand consensus one must consider a handful of issues including the character of the group, the constraints of time, the role of the facilitator, and group dynamics; all of which are made more difficult in the Wikipedia context.

Deliberation and Openness

There are numerous methods for making group decisions; one might flip a coin, vote, or seek consensus—among others. Each has its merits and difficulties, and is more appropriate to some situations than others. Unlike the first two methods, consensus is not so much about quickly yielding a "yes" or "no," but in arriving at the best possible solution. While the progress and the outcome of consensus are rarely assured, the focus is on the potential benefits of deliberation rather than the speed of the decision. (However, if consensus is achieved, the legitimacy of the decision will likely exceed that of a coin toss or vote.) As Wikipedia's "Consensus" policy notes: "Achieving consensus requires serious treatment of every group member's considered opinion. . . . In the ideal case, those who wish to take up some action want to hear those who oppose it, because they count on the fact that the ensuing debate will improve the consensus." And even though polling may be a part of the consensus process, it is "often more likely to be the start of a discussion than it is to be the end of one."[25]

But if consensus is a discussion, who is invited to the conversation? The IETF Guidelines notes: "It can be particularly challenging to gauge the level of consensus on a mailing list." A mailing list probably has many more subscribers than actual active participants, and the number of messages is not a good indicator of consensus "since one or two individuals may be generating much of the traffic."[26] Furthermore, the W3C makes allowances for notions such as quorum, supermajority, and members in good standing; these can be specified at the beginning of a group's work in its charter.[27] Wikipedia, and its topical projects, have no charter or formal list of members in good standing. It lacks many of the mechanisms other communities have to make the process of coming to consensus a little easier. Its openness is particularly problematic because it is susceptible to trolling and "forum shopping." In the case of trolling, someone who simply wants to annoy others can ensure unanimity is never achieved and increase the chances that the group will collapse in frustration. With respect to forum shopping, the consensus policy notes: "It is very easy to create the appearance of a changing consensus simply by asking again and hoping that a different and more sympathetic group of people will discuss the issue."[28] For example, during a dispute about removing articles on "marginally notable characters" it was suggested that a policy about such biographies be documented. Another Wikipedian responded: "The problem is that we have nowhere near consensus for such a policy. A large number of editors support it, and a large number of editors oppose it. Different specific cases have gone different ways, mostly depending on who showed up to the debate that day."[29] Wikipedia consensus policy counsels that this "is a poor example of changing consensus, and is antithetical to the way that Wikipedia works" and turns, again, toward reasoned deliberation: "Wikipedia's decisions are not based on the number of people who showed up and voted a particular way on a particular day; they are based on a system of good reasons."[30]

Time and Precedence

If consensus is a process whereby participants discuss and reason together, openness has another challenging implication beyond the question of who is contributing to the conversation: in an open and forever-changing group, how long might any decision be considered the group's consensus? On first blush—and beside those Wikipedia "foundation issues" considered to be beyond debate such as neutral noint of view—"consensus is not immutable."[31]

Consider a Wikipedia discussion related to how annoying it is when a bookmark or link to a Web page no longer works. Tim Berners-Lee, in the essay "Cool URIs Don't Change," writes, "Pretty much the only good reason for a document to disappear from the Web is that the company which owned the domain name went out of business or can no longer afford to keep the server running. Then why are there so many dangling links in the world? Part of it is just lack of forethought."[32] The (English) Wikipedia, understandably, suffered from such a lack of forethought when it succeeded beyond expectation and wanted a home for other language versions. What should the URI "http://www.wikipedia.com/wiki/Chernobyl," which used to identify the English article on Chernobyl, point to now? If "www.wikipedia.org" became a portal for all large language editions, and the English article was moved to an English namespace (i.e., "http://wikipedia.com/en/Chernobyl"), should the old URI "break" (i.e., return an "uncool" 404 error message) or redirect to the new location? Wikipedian Rowan Collins wrote, "I think talking of this as a 'contract' [to not break URLs] is somewhat overdoing it—it's an important point that this was the compromise reached during a previous discussion, but unless there's a *very* strong statement promising to uphold it 'forever', we generally treat all consensus policies as renegotiable."[33] This is in keeping with the consensus policy, which states, "It is reasonable, and sometimes necessary, for the community to change its mind."[34] Change is sometimes reasonable, but eternally arguing about the same thing is not, as Wikipedian Philip Sandifer notes: "I've been fighting with the same people over issues with reliable sourcing for well over a year, for instance, and yet those fights still continue despite, seemingly, a substantial shift in opinion away from the former hardline positions (things that included overbroad statements about blogs "never" being reliable sources)."[35]

As in many of the issues facing Wikipedia, Wikipedians must achieve a delicate balance, this time between rehashing tired issues and reconsidering vital ones: between "the need for open and fair consideration of the issues against the need to make forward progress." On this point, the IETF and W3C have some means for judging the merit of an issue. First, working group charters are carefully constructed so as to focus on issues that are amenable to resolution within a specified time frame. Second, the working group chair has a critical job in summarizing and recording discussion; while "it is occasionally appropriate to revisit a topic, to reevaluate alternatives or to improve the group's understanding of a relevant decision," "unnecessary repeated discussions" can be avoided with careful records of previous arguments and conclusions.[36] Additionally, reasonable criteria for the consideration of an issue are articulated by the IETF Guide:

> To facilitate making forward progress, a Working Group Chair may wish to decide to reject or defer the input from a member, based upon the following criteria: Old: The input pertains to a topic that already has been resolved and is redundant with information previously available; Minor: The input is new and pertains to a topic that has already been resolved, but it is felt to be of minor import to the existing decision; Timing: The input pertains to a topic that the working group has not yet opened for discussion; or Scope: The input is outside of the scope of the working group charter.[37]

Such criteria can be applied to not only newcomers, but also those who were present, but silent, or changed their minds. Despite all the focus on conversation and the cacophony that sometimes accompanies the consensus process, silence is one of the greatest challenges to successful decision making. In a working group, silence in response to a request for comments or objections is rarely a good thing; hopefully participants will at least say, "Sounds good to me." While one would like to think that "silence implies consent, if there is adequate exposure to the community," as Wikipedia's consensus policy states,[38] this can be a risky inference. Instead, silence often indicates confusion or a lack of interest. (On working-group teleconferences one might hear members speaking to colleagues at their office, their children at home, and even snoring!) At the IETF and W3C this prompted a step in their processes called "Last Call": before engineers begin seriously implementing and testing the specification, those who have the right to deliberate in consensus also have an obligation to make their views known. This sentiment is also famously captured in the Anglican wedding ceremony: before two newlyweds are married those who would object must "speak now; or else for ever hold your peace." Or, in the less politic words of a Wikipedian, people should "either put up or shut up": people don't have to participate, "but when they don't they should not moan when their voice is not considered."[39]

The Facilitator

In many consensus-based communities a facilitator performs a number of tasks, the most important of which is positing a consensus statement. The articulation of such a statement and asking for objections is central to Jane Mansbridge's definition of consensus in her comparative study of decision making titled *Beyond Adversary Democracy*; she uses the term *consensus* to "describe a form of decision making in which, after discussion, one or more members of the assembly sum up prevailing sentiment, and if no objections are voiced, this becomes agreed-on policy."[40] This offering of an understanding and asking for objection is also part of the answer to the question of how a group knows consensus has been reached: when there are no objections. The IETF recommends that when the chair of the working group believes she discerns a consensus she should articulate her understanding of the consensus position and ask for comments; then, "It is up to the Chair to determine if rough consensus has been reached."[41] A similar understanding exists at the W3C. However, determinations by the chairperson must be tenable to the working group and are reviewable by higher-ups, particularly after Last Call, in each institution (e.g., IETF area directors, and W3C director and advisory committee).

Among Quakers, Michael Sheeran notes that facilitators are known as humble "clerks"—reminiscent of a Wikipedia discussion equating administrators with "janitors"; nonetheless, they can wield significant influence beyond their seemingly simple responsibilities, much as is often alleged in the case of Wikipedia administrators. (On this point, Mailer Diablo's Second Law of Wikipedia is adapted from Stalin to read: "The Wikipedians who cast the votes decide nothing. The sysop/[bureau]'crat who count the votes decide everything."[42]) In his chapter on leadership, Sheeran notes, "The clerk's responsibilities" might also serve as "devices for hidden control" with respect to setting the agenda (i.e., scheduling which issues are discussed and when), stating questions (i.e., in an even-handed manner), facilitating the discussion (i.e., encouraging participation and discouraging obstructionists), judging what is important (i.e., whether something is substantive or trivial), and judging the sense of the meeting (i.e., is there consensus?).[43] These devices might be used for good (e.g., structuring the agenda so that a working group gains momentum from easier issues) or ill (e.g., scheduling an unwelcome item at the end of a full agenda).

These communities benefit from the (hopefully) wise guidance of a trusted community member, be it a working group "chair" or meeting "clerk." In much of Wikipedia decision making there is rarely any such formally identified resource at the start. (Though through requests for comments, mediation, or arbitration such a person might become involved.) The effects of this can be seen in the naming controversy with which I opened the chapter. Wikipedian Wknight94 summarized the issue in his statement this way:

> A clear-cut case of supermajority consensus has become a nasty all-out war with a very vocal minority. A poll which is now visible here included a question of whether television episode articles should only be disambiguated when necessary. . . . The result was 26 people choosing to support disambiguating only when necessary and seven choosing to oppose. The poll was well-publicized. Nonetheless, a few members of the minority, mostly Elonka and occasionally MatthewFenton, have declared that there was no consensus and that the dispute is still open. The reason most often given is that the poll was modified several times while in progress. While that is true, it was mostly modified from a one-question poll with three choices to a two-question poll, each with two choices, and the meaning of the most contentious issue remained unchanged (not to mention Elonka herself modified the poll. . .).[44]

Wikipedian Josiah Rowe spoke of a possible confusion about the meaning of consensus and the seeming triviality of the issue:

> Of course, consensus does not mean unanimity, but as long as we were short of unanimity, Elonka (and one or two others) insisted that the poll needed to be re-run. . . . The core issue of this debate, how to name articles about television episodes, is really quite unimportant in the greater scheme of Wikipedia. I really don't understand why the debate got to this point, and it saddens me that it has. Any resolution would be welcome.[45]

Beyond affirming the principle that one should abstain from personal attacks—particularly of a sexual nature—the ArbCom responded to this dilemma by focusing on the failure to "close" the discussion.[46] This focus on closure is puzzling in that most users believed the case was about naming policy and "about consensus—whether it was reached." Wikipedian Yaksha continued: "The result i'm hoping for is just a declaration that we got consensus, and that people should respect consensus."[47] Yet the ArbCom characterized the issue as a procedural one, as noted in the affirmed principles of the decision:

> 1.3) After extended discussion, to be effective, the consensus decision making process must close. . . . In other, less structured, situations, as in the case of how to structure the titles of television episodes, there is no formal closer. Nevertheless, considering the alternatives proposed, the extended discussion engaged in, expressions of preference, there is a result which should be respected. Absent formal closing, it is the responsibility of users to evaluate the process and draw appropriate conclusions.[48]

This is somewhat surprising and confusing. One must look closely in the decision's "finding of fact" to see that consensus is presumed by the ArbCom ("a consensus decision was reached") without stating how this conclusion is arrived at, and most importantly, how Wikipedians can convince a recalcitrant minority when this is the case. As Yaksha noted, he believed there was consensus and began to rename articles to disambiguate only when necessary, but "Elonka, however, claimed that there was no consensus, to move the articles, and that the moves were disruptive." And so the "edit war" began. Wikipedians found the decision ultimately unsatisfying, and in the following thread Wikipedian badleydrawnjeff asked who was responsible for knowing when to close, and ArbCom member Fred Bauder attempted to respond:

> The final decision notes that "It is the responsibility of the administrators and other responsible parties to close extended policy discussions they are involved in." What is a "responsible party?" What sort of expectation is it to close an "extended policy

discussion?" At what point is it "extended," and at what stage is it okay to throw in the towel? At an arbitrary moment or simply when the discussion becomes "disruptive?" Thanks. —badlydrawnjeff 22:15, 21 January 2007 (UTC)

An established and respected user who is not an administrator could close a discussion. An extended policy discussion is one in which most aspects of the question ha[ve] been discussed, alternatives considered, in short, a full discussion. Good judgement is needed to determine when consensus has been reached or when it is obvious there is no consensus. When the discussion becomes disruptive, more heat than light, it is probably past time to close the discussion and declare a result. Fred Bauder 22:38, 22 January 2007 (UTC)

So nothing really specific, per se? —badlydrawnjeff 01:21, 23 January 2007 (UTC)

The subject does not lend itself to bright line rules. The question is whether the question has been fully discussed and a decision reached. Fred Bauder 01:52, 23 January 2007 (UTC)

However, it is still difficult to see how this situation could have been avoided. Wikipedian Nedd Scott joined the conversation by noting that Bauer's response wasn't really useful, resulting in this exchange:

It's basically saying "If you think you're right then say so and tell everyone to shut up." Won't everyone think they're right in a discussion/dispute/etc? If the situation is reasonably clear one way or the other then we usually don't have to resort to something like this to end it. The situations this is supposed to be helpful in are usually too unclear to actually use this.—Ned Scott 05:24, 24 January 2007 (UTC)

Wikipedia:Requests_for_arbitration/Naming_Conventions involved a matter where there was a consensus, but no closing. Based on lack of closing, an opposition party engaged in move warring. That was the problem we were trying to address. Fred Bauder 03:10, 25 January 2007 (UTC)

I guess that's one way to look at it, but the solution offered still isn't helpful. Nothing personal. —Ned Scott 04:27, 25 January 2007 (UTC)

Fortunately, Wikipedian Ace Class Shadow offered some useful advice while recognizing there is no perfect solution: one could use templates to mark that a discussion page is now archived, or one could "ask an impartial closer to do the deed, stating that you'll respect their common sense judgment. To this day, I've only encounter[ed] one closing that, using this method, seemed at all inaccurate."

For this dispute, it is not clear if this guidance could have convinced the minority supporting consistent-suffix use of the legitimacy of the "use as needed" policy. Therefore, beyond the censure on personal attacks and "given the existence of some uncertainty regarding how to determine if there is consensus in a particular case," no punishment was proposed on any of the parties to the case "for past violations of policy."[49] Such is the ambiguity and challenge of consensus practice, and a possible source of temptation to use some system of voting.

Polling and Voting

Consensus is the preferred method of making decisions at Wikipedia. This is as much because of this method's merits (e.g., discovering mutually beneficial solutions) as its alternatives' demerits. While consensus can be difficult, Wikipedians frequently cite the aphorism that "Voting Is Evil."[50]

Yet, as seen earlier, polling is an available technique within the consensus process. When a poll is taken on Wikipedia, individuals are invited to list their position under one of the specified options (e.g., A or B; accept, reject, abstain) with an explanation, which then might prompt further commentary and discussion. How is polling different from voting? While people may confuse polling with voting—or even speak of voting as "a quick shorthand for what we are actually doing"[51]—polling should prompt and shape discussion, rather than terminate it:

> Wikipedia operates on discussion-driven consensus, and can therefore be regarded as "not a democracy" since a vote might run counter to these ends. Some therefore advocate avoiding votes wherever possible. In general, only long-running disputes should be the subject of a poll. Even then, participants in the dispute should understand that the poll does not create a consensus. At best, it might reflect how close those involved are to one.[52]

In fact, even polling is considered suspect—and "evil"—by some as it is thought to discourage consensus, encourage groupthink, be unfair, be misleading, and encourage confusion;[53] the botched poll in the disambiguation case is evidence enough of possible pitfalls.

To be fair, consensus doesn't work well in all circumstances. It is best suited to small groups of people with some common interests and acting in and assuming good faith. It requires a community, not just an electorate. Jane Mansbridge, in her study of decision making, finds that groups with the largest number of interdependent friendships were those most likely to achieve a consensus that "did not paper over an underlying divided vote."[54] But as a group grows the "Community May Not Scale," as is noted at Meatball (the wiki about wiki collaboration). Meatball's consensus page states that as the size of the group (n) increases, so does the chance of conflict between individuals (n^2) and between subgroups (2^n):

> Voting is one of the best ways to quantify opinions in a large group. Online communities that have a common goal will be continually in need of making decisions. In a small group of similar-minded individuals, a consensus decision can often be found by discussion. In such a group, VotingIsEvil.
>
> However, as the group grows larger, CommunityMayNotScale. For each member of the group, there's a certain likelihood that he disagrees with one of the existing members. The likelihood of conflict between two individuals grows geometrically, according to MetcalfesLaw. The likelihood of conflict between two sub-groups grows exponentially, according to ReedsLaw. If individuals with contrary opinions are CommunityExiled, the group may succumb to GroupThink. If not excluded, they may delay consensus decisions indefinitely. Therefore, VotingIsGood.[55]

Clay Shirky writes that social software must be designed so as to protect a group from becoming "its own worst enemy" by finding a way "to spare the group from scale."[56] In fact, theorists and members of online community alike cite "Dunbar's number" or "the Rule of 150" to indicate the challenges of community growth. (This idea of a maximum limit on the number of stable interpersonal relationships that we can maintain was popularized by Malcolm Gladwell in *The Tipping Point*; he gives many examples including the Hutterites, a communal branch of Anabaptists, who split a colony in two once it reaches 150 members.[57])

Additionally, consensus is not a panacea for the difficulties inherent to group decision making. The Wikipedia "Consensus" article notes that consensus can take a long time, be frustrating in circumstances where there is little hope of agreement, and, when understood as unanimity, can give a self-interested minority veto power over group decisions.[58] Furthermore, the "Consensus Decision-Making" article notes that consensus is inherently conservative (i.e., preserves the status quo), susceptible to disruption, and can give rise to groupthink by which members suppress their own opinions for the sake of conformity or group harmony. This can even yield a paradox in which the group's final position is held by few members, such as when members falsely, for harmony's sake, support a "cascading" but minority position because it benefited from being expressed first.[59]

And while voting may be appropriate in some circumstances, or at least a last resort if consensus fails, the openness of Wikipedia, again, contributes to the sentiment that voting "is evil." Meatball notes "online voting suffers badly" because people can "stuff the ballot box" or bias the framing of the poll.[60] The "sock puppet," a cousin of the "troll," is an account used to "create the illusion of greater support for an issue, to mislead others, to artificially stir up controversy, to aid in disruption, or to circumvent a block."[61] (Think of a literal sock puppet on your hand, agreeing with everything you say.) These types of problems are particularly prominent in the voting associated with "Articles for Deletion" (AfD, where Deletionists and Inclusionists duel) and "Requests for Adminship" (RfA, where bitter rivalries flourish).[62]

While the distinction between polling and voting might blur as an argument grinds on and the possibility of consensus declines, the spirit and mechanics of the two are different. Consider that in consensus, if a group has not succumbed to groupthink, one should feel free to speak of and identify with a minority position. People will need to know whom to engage with and potentially learn from. Yet in democratic voting, the secret ballot has significant advantages. In a thread about a resolution of the Wikimedia Foundation Board, Danny Wool, a foundation employee, wrote that he found questions about who voted for or against a resolution troubling: "In a true democratic system, the secret ballot allows people to vote their conscience, rather than voting for popularity, material reward, fear of censure, and whatnot. A commitment to openness should not be misused so cynically."[63] Additionally, consensus presumes good faith and sometimes sustains it; voting can operate without good faith and sometimes depletes it altogether. Mansbridge argues the differences between voting and consensus can be understood thus: "Voting symbolizes, reinforces, and institutionalizes division. . . . while a decision by consensus includes everyone, reinforcing the unity of the group."[64] This can also be seen in different Internet standards organizations. Granted, "politics" are present in every venue, but in consensus organizations participants might be more likely to show up with an open mind and willingness to engage others. In voting-based institutions, "stuffing the ballot box" can get out of hand. This was demonstrated most effectively in a case that merits a brief digression.

In response to the growing popularity of the OpenDocument Format (ODF) standard, an alternative to the proprietary format used by MS Word, Microsoft offered a new version of its formats through a "fast-track" international standard process.[65] For supporters of open standards this could be a significant gain except that the format is said to be complex and vendor-biased and that it does not make full use of other standards, and is difficult to fully implement.[66] When it came time to vote Microsoft is alleged to have pressed allies to acquire voting ("P") memberships in an International Organization for Standardization (ISO) working group for the purpose of approving the format (i.e., identified as ECMA 376 / ISO 29500). It appears the standardization effort has been successful, although some claim the voting process was abused.[67] And there has been an interesting—and unfortunate—side effect. In one of his final reports as chair of the responsible working group, Martin Bryan noted, "The second half of 2007 has been an extremely trying time for WG1. I am more than a little glad my 3 year term is up, and must commiserate with my successor on taking over an almost impossible task." In addition to the short time frame, and the interdependencies and complexity of the task, the arrival of otherwise uninterested members made subsequent work impossible to complete because those who joined to vote on this single issue subsequently disappeared, preventing a quorum:

> As ISO require[s] at least 50% of P members to vote before they start to count the votes we have had to reballot standards that should have been passed and completed their publication stages at Kyoto. . . . The days of open standards development are fast disappearing. Instead we are getting "standardization by corporation," something I have been fighting against for the 20 years I have served on ISO committees. I am glad to be retiring before the situation becomes impossible. I wish my colleagues every success for their future efforts, which I sincerely hope will not prove to be as wasted as I fear they could be.[68]

At Wikipedia, additional confusion arises about the gray area between consensus and voting. In voting systems, it is not uncommon for like-minded individuals to advocate, campaign, and horse trade (i.e., exchanging votes on different issues).[69] In a thread on the English Wikipedia list, Wikipedian Johntex advocated that "We Need to Recognize That Advocating Is a Basic Right"; he argued users should be able to "influence policy in ways that they believe are beneficial to the project," including "building up groups of people who agree with you and who will help you bring about the beneficial change." Advocating for a cause or recruiting those who already agree with you is "a Good Thing":

> Let's stop insulting people by calling them "meat puppets" or "vote stackers." Let's stop confusing the issue by calling it "spamming." It is not spamming. Spamming is indiscriminately notifying people that are probably not interested in the hopes that a few people will be. This is practically the opposite.
>
> Attempting to stifle advocacy is harmful to the consensus building process and it is harmful to the project. If we try to prohibit it, it will just be taken off-wiki, which would be a huge shame.[70]

But others disagreed, finding these practices from the democratic sphere to be counter to consensus practice. Wikipedian Tony Sidaway responded, "We don't do advocacy or campaigning on Wikipedia. Our decision-making processes are deliberative rather than democratic."[71] Yet, the difficulties of achieving consensus and this blurring of approaches to decision making prompted Zoney to declare that any pretense of consensus should be abandoned and the community should move toward a "fixed method of decision-making":

My problem is that perpetuating the lie that decision-making on Wikipedia is by consensus, we don't strictly adhere to any other decision-making form (e.g. majority voting). In consequence, decisions are "whatever people can get away with." Of course if there's an actual real consensus (general agreement) then there's a valid reason for a decision not being challenged. But more often than not all it means is that influential individuals ensure they get their way and others give up (that isn't forming consensus by the way), or else we have majority/mob rule.[72]

In the resulting thread some conceded there are problems, but not as bad as feared. The discussion touched on many of the challenges discussed in this chapter. And some took a pragmatic "time will tell" approach: hopefully the right thing happens more often than not. Sidaway felt that ultimately, "Decisions are more likely closer to 'whatever offends the least number of people'. This is sometimes less than optimal—I could give my list of things I think are poor decisions and you could probably give yours. The result is that nobody is ecstatic but we have something we can move ahead with."[73] Likening emerging consensus to nailing jelly to the wall and keeping the stuff that sticks, he continued his pragmatic perspective: "If hundreds of people edit a piece of work in good faith over a long period, what changes least over time may be presumed to be there by consensus. However even the most apparently stable elements of a work may be deposed quite easily. The result may be a new consensus or, in other cases, a period of instability where the new version and the old version compete."[74] In response to the discussion Zoney asked in frustration if there is then "any hope of having a fixed method of decision-making on Wikipedia, rather than a shambolic pretence of achieving consensus . . .?" Wikipedian Adrian responded simply, "No."[75] Marc Riddell responded a little more encouragingly, "Yes, there is hope; if we can put our individual egos and emotions aside—and start using our heads in a responsible way."[76]

Conclusion

Humans naturally look for means by which difficulties can be clearly dissected and neatly dispatched. Yet, given its reliance upon an assumption of good faith and a preference for consensus in its decision making, one can conclude that the Wikipedia community is relatively tolerant of the ambiguities inherent to collaborating on a world encyclopedia and rather trusting of human judgment over the long run. While Wikipedia must often address many of the conflicts present in the real world, and has plenty unique to its own mission and methods, I would never argue that Wikipedia has become the global institution of "adjustment and adjudication" that Wells foresaw. However, Wikipedia is a fascinating example of a historic means of community decision making in a new context. In particular, its openness—the lack of topical and temporal scope, the initial lack of facilitation, the turnover in membership, and anonymity—brings a new salience to the challenges of consensus practice.

These challenges are seen in the humorous—perhaps cynical—definition of consensus at the head of this chapter (i.e., the scope of consensus includes only those who share your opinion). Similarly, in response to the question "How many Wikipedians does it take to change a light bulb?" Wikipedian Durova answered "69":

> 1 to propose the change; 5 to support; 1 to dispute whether the change is a needed process; 7 more to pile on from IRC [Internet Relay Chat] and join the dispute; 2 to open a request for comment; 37 to vote at the straw poll; 5 to say votes are evil; 1 to MFD it [propose it as "Miscellany For Deletion"]; 9 to object until the MFD gets speedily closed; 1 to mark the proposal historical. Afterward on AN [Administrators' Noticeboard], all opposers claim the consensus favored darkness [i.e., no light bulb!].[77]

In another WikiSpeak entry *consensus* is defined as "one of the three states that can be reached at the end of a discussion after all parties have become thoroughly fed up with it; the alternatives are no consensus or for pity's sake, I wish I'd never gotten involved in this."[78] It can be an altogether frustrating experience. And in some circumstances, such as irreconcilable differences between community members or external threat, this tolerance can be incapacitating. And so, Wikipedia, like other open content communities, is also characterized by an odd type of leadership at the highest level: the "benevolent dictator," the subject of the next chapter.

6 The Benevolent Dictator

When a building is on fire, a leader will not survey everyone to see what the consensus is about a response. It is time for action.
—Bhadani's Second Law

Open, civil, egalitarian, deliberative: these are some of the concepts encountered in the pursuit of a universal encyclopedia. While they might seem simple enough in the abstract, they become much less so when used in the practice and discourse of a community. For instance, a perfectly "open" community will likely be chaotic, rendering it inhospitable to many. Or, if consensus doesn't require unanimity, agreement—unanimous or otherwise—on what it does require can be elusive. Some of the sources and ironies of the English Wikipedia's collaborative culture are further highlighted when one considers the role and status of leadership. Wikipedia, like other open content communities, is predominately a voluntary effort—aside from a few Wikimedia Foundation staff—and there's little room for coercion or utilitarian rewards.[1] Yet there is often a seemingly paradoxical use of informal tyrant-like titles (i.e., "benevolent dictator") for the community leader. What, then, can we make of this latest puzzle?

In this chapter I show how this juxtaposition can be understood as an "authorial" form of leadership whereby exceptional autocratic power is exercised by a respected "author" within an open content community. I then return to the story of Wales and Sanger, for their conceptions of leadership and expectations for the community profoundly shaped its direction and culture. Finally, I consider how the community discusses this type of leadership and the values with which it seems at odds.

Authorial Leadership

During one of the discussions about a feared "neo-Nazi" attack with which I began this book, Jimmy Wales responded, "If 300 NeoNazis show up and start doing serious damage to a bunch of articles, we don't need to have 300 separate ArbCom cases and a nightmare that drags on for weeks. I'll just do something to lock those articles down somehow, ban a bunch of people, and protect our reputation and integrity."[2] How can such an autocratic statement be made within a supposedly open and consensus-based community? (I continue to use the term *autocratic* to describe, nondisparagingly, leadership actions that do not derive their authority from group decision-making processes.) Actually, such an exercise of power by a community founder is not unique to Wikipedia. Such "authorial" leadership is common to many open content communities and prompts three questions that merit attention: What is the environment from which such leadership emerges? How is it enacted? And, most interesting, how is it discussed and understood by the community?[3]

With respect to the environment, such leaders often found a project (often by authoring the initial content) around which a community develops, or otherwise emerge from a leaderless context by way of merit; subsequently they influence the direction of a community's culture.[4] Furthermore, this type of leadership often operates within a mix of governance models: meritocratic (setting the direction by leading the way), autocratic (acting as an arbiter or defender of last resort), anarchic (consensus); and occasionally democratic (voting).[5] Wales himself has noted that:

> Wikipedia is not an anarchy, though it has anarchistic features. Wikipedia is not a democracy, though it has democratic features. Wikipedia is not an aristocracy, though it has aristocratic features. Wikipedia is not a monarchy, though it has monarchical features.[6]

With respect to conduct, leaders often convince by persuasion and example though they also retain charismatic authority accumulated from their merit.[7] This authority is frequently employed to act, as a last resort, as an arbiter between those of good faith or as a defender (but an autocratic one) against those of bad faith. As Free and Open Source Software (FOSS) luminary Eric Raymond notes, leaders must be capable of operating with a "soft touch," to "speak softly," consult with peers, and "not lightly interfere with or reverse decisions" made by other prominent members.[8] Additionally, humor and civility facilitate camaraderie between all participants and ease the exercise of authority and related anxiety.

Finally, such leadership is rarely enacted or understood as a formal office, though prominent leaders might be endowed with the informal moniker of "benevolent dictator" and occasionally act autocratically,[9] as Wales threatened in the neo-Nazi case. However, leaders whose autocratic actions exceed their accumulated merit or charisma, sometimes referred to "idiosyncrasy credits" or "reputation shares," risk their status and a forking of the community.[10] For example, while a "benevolent dictator" might be tolerated as a necessity, a "God King" is a "site owner or administrator who uses their special authority more than absolutely necessary." This is a leader so "arrogant that they suppose they are 'god'"; this type of leadership is an "abuse," "a bad thing," and an "anti-pattern" of good wiki community.[11] Also, the possibility of forking—even if unlikely—is central to voluntary community dynamics and discourse, as David Wheeler notes with respect to FOSS communities:

> Fundamentally, the ability to create a fork forces project leaders to pay attention to their constituencies. Even if an OSS/FS project completely dominates its market niche, there is always a potential competitor to that project: a fork of the project. Often, the threat of a fork is enough to cause project leaders to pay attention to some issues they had ignored before, should those issues actually be important. In the end, forking is an escape valve that allows those who are dissatisfied with the project's current leadership to show whether or not their alternative is better.[12]

In short, only those leaders that tread carefully and continue to make important contributions (including, now, the judicious exercise of autocratic authority) are granted the "dictator" title. Whereas this term might not be the most appropriate in capturing the genuine character of this role, it serves as a warning: a good-natured joke balanced on the edge of becoming a feared reality.[13] It serves as a caution to such leaders, as well as a metaphoric yardstick for discussing any participant's action.

Because of the voluntary and meritocratic character of open content communities it is not surprising that leaders are expected to lead by example as their very leadership is founded upon exemplary behavior; leadership emerges through action rather than appointment. And while a founding leadership role has some semblance of authoritarianism to it, at least in title, it is eternally contingent: a dissatisfied community, or some constituency thereof, can always leave and start again under new leadership.

Wales and Sanger

Two of the most influential people in the history of Wikipedia are cofounders Larry Sanger and Jimmy Wales. In *Organizational Culture and Leadership,* Edgar Schein identifies ways in which such leaders embed and transmit culture including "how leaders react to critical incidents and organizational crises."[14] The following brief account of the crisis of Nupedia's demise, Wikipedia's rise, and Sanger's departure provides a revealing introduction to leadership in the Wikipedia context.

Wales, a co-owner of the Internet content and search company Bomis, hired Sanger in February 2000 to launch and act as the editor in chief of the Nupedia project. Until he resigned, Sanger was the most prominent leader of Nupedia (the original peer-review project) and Wikipedia (its wiki complement and eventual successor). As Sanger writes in his April 2005 memoir:

> The idea of adapting wiki technology to the task of building an encyclopedia was mine, and my main job in 2001 was managing and developing the community and the rules according to which Wikipedia was run. Jimmy's role, at first, was one of broad vision and oversight; this was the management style he preferred, at least as long as I was involved. But, again, credit goes to Jimmy alone for getting Bomis to invest in the project, and for providing broad oversight of the fantastic and world-changing project of an open content, collaboratively-built encyclopedia. Credit also of course goes to him for overseeing its development after I left, and guiding it to the success that it is today.

What precipitated Sanger's resignation? As discussed in chapter 2, Sanger was caught between continuing frustration with Nupedia's slow progress on one hand and problems with unruly Wikipedians on the other. Furthermore, Sanger alienated some Wikipedians who saw his actions as unjustifiably autocratic and he eventually broke with the project altogether. In late 2006 Sanger launched the more expert-friendly collaborative encyclopedia Citizendium. In any case, Sanger's account recognizes the uneasy tension between title and authority and cultural momentum at the founding of this community:

> My early rejection of any enforcement authority, my attempt to portray myself and behave as just another user who happened to have some special moral authority in the project, and my rejection of rules—these were all clearly mistakes on my part. They did, I think, help the project get off the ground; but I really needed a more

subtle and forward-looking understanding of how an extremely open, decentralized project might work.

Such an understanding might have been like that of Theodore Roosevelt's recommended leadership style: speak softly and carry a big stick. Whereas Sanger did have special authority at Nupedia as editor in chief, such was not the case at Wikipedia, and Sanger's corresponding "loudness" was a later cause of regret:

As it turns out, it was Jimmy who spoke softly and carried the big stick; he first exercised "enforcement authority." Since he was relatively silent throughout these controversies, he was the "good cop," and I was the "bad cop": that, in fact, is precisely how he (privately) described our relationship. Eventually, I became sick of this arrangement. Because Jimmy had remained relatively toward the background in the early days of the project, and showed that he *was* willing to exercise enforcement authority upon occasion, he was never so ripe for attack as I was.[15]

Perhaps unrealized by Sanger, Wales exhibited this pattern of leadership even on an earlier philosophical email list, for which he wrote that he would "frown *very much* on any flaming of any kind whatsoever" and choose "a 'middle-ground' method of moderation, a sort of behind-the-scenes prodding."[16] And most interestingly, Sanger attributes a root of the problem to his failure to recognize the importance of community and culture:

For months I denied that Wikipedia was a community, claiming that it was, instead, only an encyclopedia project, and that there should not be any serious governance problems if people would simply stick to the task of making an encyclopedia. This was strictly wishful thinking. In fact, Wikipedia was from the beginning and is both a community and an encyclopedia project.[17]

As noted earlier, upon publication of Sanger's memoirs a controversy arose over whether Sanger even deserved credit as a cofounder of Wikipedia. In a sense, in playing the bad cop one is depleting one's own reputation or leadership credits in favor of the good cop; Sanger, in shifting from bad cop to apostate, prompted some to question whether such credit was merited at any time. A more productive discussion at the time characterized the change in leadership style as a necessary one:

Now, I must say. . . I think a project of such a type can only work *without* a strong authority. It is important to let people built their own organisation. Jimbo has this very powerful strength, in this that he lets most of the organisation be a self-organisation. For those who know a bit about leadership, it is a rather rare occurrence. For the sake of wikipedia, and to let all the international projects grow up (without a strong hand to lead them), it was important that the role of the editor in chief disappear.[18]

Sanger actually concedes as much in the development of editorial policies but is still concerned about controlling abusive editors and attacks, particularly when they alienate high-quality expert contributors. And so he now leads the Citizendium project.

Wales's Influence

Authorial leaders are frequently the initial author of the community's content. This is the case, for example, with Linus Torvalds and the Linux kernel or Guido van Rossum and the Python programming language. In this respect, Wikipedia is a bit different, as was pointed out to me by Evan Prodromou, Wikipedian and a founder of Wikitravel.[19] Prodromou argued that unlike FOSS communities, Wikipedia has many more contributors, many of whom, even at the administrator level, contribute at a low skill and intensity level compared to FOSS contributors. Furthermore, unlike other wiki communities or even other leaders within Wikipedia, Wales has never been a significant "author" in terms of creating content. Indeed, because of Wikipedia's history, the community regards an editor in chief as undesirable, and even Wales's relatively modest editorial contributions are apt to cause concern. (In fact, in *The New Yorker* he admitted he abandoned his efforts to write a scholarly Nupedia article on Robert Merton and options-pricing theory because it was too intimidating and reminiscent of graduate school.[20] Sometimes his Wikipedia edits are challenged, as we will see, and statistics on his contributions and "edit count" have been a topic of discussion.[21]) Plus, much of his purview has been understandably limited to English projects. And even though Wales's public presence in the daily life of Wikipedia has receded,[22] I consider his leadership to be central because of his founding vision, early activity, contributions to collaborative norms, relationships with other Wikipedians, and latent power.

In addition to reacting to crises, Schein argues that community culture is affected by what leaders "pay attention to."[23] In this way, leadership can be exerted by highlighting rather than coercing. For example, in any early discussion about neutral point of view, Wales identifies an important issue and highlights a sentiment he agrees with: "We should all pay close attention to Larry's wording here, which I think is excellent. Nupedia should 'include articles *on* all points of view' (note the emphasis added), not necessarily 'include articles *from* all points of view.'"[24] Or, as seen in the discussion about the blocking of a white supremacist, Wales went out of his way to commend the participants for having "a disagreement with a positive exploration of the deeper issues."[25] Highlighting others' arguments to make his own has even led Wales to apologize for contravening Netiquette; in a thread about the balance between high-quality content and "cruft," Wales commented: "I know it is bad form to quote an entire post just to say 'me too' but I wanted to say that Daniel is right on the money here, and displays what I think of as true Wikipedia spirit. We have to have a passion to *get it right* or we'll be full of rampant nonsense."[26] He also can be found highlighting what he thinks to be central Wikipedia values: "Wikipedia is built on (among other things) twin pillars of trust and tolerance. . . . The harmony of our work depends on human understanding and forgiveness of errors."[27]

Furthermore, after immersing oneself in Wikipedia practice it is not difficult to see that many of its good faith norms are strongly exercised by Wales himself. In a 2007 discussion about his role at Wikipedia he described his approach as diplomatic and reflects elements of both good faith and neutrality:

> I have many faults, but refusal to listen is not really among them. I make mistakes, but I am calm and educable. I try to land in the center on most issues, rather than staking out any sort of extreme positions. And I try to represent all parts of the community's interest in the broad building of consensus as being better than gang warfare.[28]

Wales once described his approach to me as "I like to think I'm not stupid, but I'm not in my present position because I'm smart but because I'm friendly."[29] This attitude can be seen in the following interactions in which Wales frequently writes with:

- patience: on a thread regarding Serbo-Croatian dialects: "For those who find Mark irritating, and who may not tend to listen to him on those grounds, I would like to say, listen to him on this point."[30]
- civility: in response to someone who spoke of a threatened fork over a Friulian dialect and challenged "ARE YOU CRAZY!?!!!!?!!?!?!" Wales responded, "Good luck with that. 'Not yelling at people' is a critical trait of leadership in an all volunteer project."[31]

• humility: in response to someone concerned about perennial problems, including language policies, Wales wrote, "I'm very sympathetic to all these points. I don't have an easy answer what to do."[32]

• a willingness to apologize: when Wales recommended some text be added to a page when it was already present he wrote, "Ok, my mistake, I'm very very sorry. I didn't see that. I apologize for any confusion."[33]

Additionally, humor serves to further camaraderie and diffuse anxiety about leadership. In response to a message about an April Fool's Day joke about Wales as dictator, someone responded that many prominent Wikipedians make jokes:

> These jokes don't have a "point." If you scour the list for all messages, you will find that I am not the only one who has a sense of humour and knows how to make jokes. In fact, this extends to Ant, Mav, Jimbo, etc. who can occasionally be found to be making a joke on this list.
>
> I don't know how it is with you, but as far as I know the point of humour is to lighten up a situation, and only occasionally to make a point.[34]

However, as Wikipedia has grown, attempts at humor by those in positions of authority seemingly become rarer because a bad or misunderstood joke can have deleterious consequences exceeding the value of a few chuckles. And, of course, just as Wikipedia sometimes fall short of its ideals, Wales—and other leaders—make their fair share of mistakes, some of which are widely publicized because of Wikipedia's prominence and a counter-culture of message boards that thrive on complaint and conspiracy.[35] Even within the community, his attempts to steer Wikipedia are sometimes challenged. For example, during the 2006 Wikimedia board of trustee elections, Wales's message encouraging people to vote—and for specific candidates—was thought inappropriate by some because he might have access to the intermediate results; subsequent elections were hosted and overseen by an external organization.[36] Or, in response to an embarrassing instance of vandalism in 2009, Wales called upon the foundation to enable the experimental "Flagged Revisions" feature at the English Wikipedia based on his "personal recommendation" and community consensus (roughly 60 percent of those polled supported the idea). This prompted a maelstrom of discussion, and mainstream press attention, about openness, the meaning of consensus, and his role. However, despite initially overreaching, he and the community continued substantive discussion and Wales challenged those who objected for a specific counter-proposal within a limited time frame.[37]

(I expect using this feature as a way to protect specific pages eventually will be implemented.)

In any case, Wikipedia's good faith culture undeniably has been shaped by Wales's own values and actions; while he did not write many articles, he did help establish many of Wikipedia's essential values and norms. Additionally, after Sanger's departure he once again attempted to move to the "background" in encouraging other forms of governance to emerge and by supporting like-minded persons with a similar temperament.

Beyond the Founders: Admins, ArbCom, and the Board

Whereas cofounder Larry Sanger was editor in chief of Nupedia and he was informally known as the chief organizer of Wikipedia, neither role was ever claimed again after he resigned. Instead, the "Administrators" page stresses that everyone is an equal editor. Those who demonstrate themselves to be good editors may request extra responsibilities but "are not imbued with special authority."[38] Yet, while Wikipedia culture values editorial egalitarianism over administrative responsibilities, this does not mean there are no leaders. Consequently, before turning to how the community speaks about leadership, I first present a brief description of the leadership and governance structure of Wikipedia itself.

A novel characteristic of Wikipedia is that most anyone who browses Wikipedia may edit it—though a tiny fraction of pages are "protected" if they are subject to persistent or severe policy violations, such as edit warring, vandalism, defamation, or copyright violations.[39] Wikipedia pages claim that contributors who sign up for an account and log in—no longer "anonymous"—do not gain additional powers; instead, they have access to useful features such as a user page and the ability to track the pages one cares about. (Of course, a logged-in user who builds a good reputation can garner informal authority among other contributors.) Additional features are made accessible to experienced users in the role of a *system administrator*, or sysop. These features permit such an administrator to enact Wikipedia policy and group consensus, particularly with respect to the management of protected pages, the deletion of pages, or temporarily blocking sources of vandalism. Yet, the English Wikipedia's "Administrators" page quotes Jimmy Wales as saying, "This should not be a big deal." Indeed, in a 2005 version of this page an association with editorial authority is purposely disavowed:

Administrators are not imbued with any special authority, and are equal to everybody else in terms of editorial responsibility. Some Wikipedians consider the terms "Sysop" and "Administrator" to be misnomers, as they just indicate Wikipedia users who have had performance- and security-based restrictions on several features lifted because they seemed like trustworthy folks and asked nicely. However, administrators do not have any special power over other users other than applying decisions made by all users.

In the early days of Wikipedia all users acted as administrators and in principle they still should. Any user can behave as if they are an administrator, provided that they do not falsely claim to be one, even if they have not been given the extra administrative functions. Users doing so are more likely to be nominated as full administrators by members of the community and more likely to be chosen when they are finally nominated.[40]

Essentially, administrators are able to quickly prevent and intervene in destructive edits. (Textual vandalism isn't truly destructive as the previous versions are available; one administrative feature is the *rollback* that permits the quick reversion of such edits.) However, in an ironic testament that administrators are much like ordinary users, they do sometimes become involved in *wheel wars*, a term going back to the 1970s to describe conflicts among those who gained "big wheel" privileges on a computer system.[41] And, given there are now thousands of contributors, administrators, and administrative actions it is no longer possible to claim that administrators are "applying decisions made by all users" as was claimed in 2005. A clarification in 2008 states: "There is very little extra decision-making ability that goes along with adminship, and it does not add any extra voice in consensus decisions. In that sense, whether a person is an administrator is not, in and of itself, important."[42]

In the time since its founding, additional levels of authority have appeared as Wikipedia evolved from a small English-only encyclopedia to a massive project among many at a nonprofit foundation. At the English Wikipedia there are now 900-plus active administrators and about a dozen active *bureaucrats* who appoint administrators and other bureaucrats.[43] Elected *stewards* can, respectively, change any such role across all Wikimedia wikis and act as bureaucrats for smaller projects.[44] Orthogonal to administrative and governance roles there are also *developers*, those who actually write the software and administer the servers.[45] Volunteers continue to act in all of these capacities: the Wikimedia Foundation has only a handful of employees who administer the foundation, solicit funding, or perform essential hardware/software maintenance and development.[46]

In Wikipedia culture, and in keeping with the larger wiki culture, delineations of authority are suspect, as is seen in the previous excerpt regarding the role of administrators. Yet, even if these other levels of authority entail responsibilities rather than rights—which is the orthodox line—they could nonetheless be seen as something to achieve or envy if only for symbolic status. This leads to the occasional call for the label associated with this role to be deprecated, as discussed in the thread "Rename Admins to Janitors":

> I'm sick and tired of people misunderstanding what an "administrator" of Wikipedia is. It was a misnomer to begin with, and we've had nothing but trouble with this name ever since. Users misunderstand it (and ask admins to make editorial decisions). Media misunderstand it (and either do not explain it, or connect it to power and influence). And it's no wonder. "Administrator" could refer to a manager, or someone appointed by a court; it typically describes someone in an important official position.
>
> When the role of "bureaucrat" was created, the name was chosen specifically so that people would not treat it as a status symbol. It should be something nobody really _wants_—something people do because it needs doing, not because it gains them credibility and influence. This seems to have worked reasonably well for the most part.[47]

Also, it is worthwhile to note that as one ascends the hierarchy of roles, and the power of implementation increases, policy discretion often decreases. Just as administrators ought not to have extra authority in making editorial decisions, stewards should not make policy decisions. Stewards can "remove arbitrary user access levels" on any Wikimedia wiki. They can toggle whether one has the ability of an administrator (to block users or protect pages), a bot (to run automatic tools), or a bureaucrat (to set access levels within a single wiki), and whether one has the ability of oversight (to suppress revisions), or checkuser (to determine the Internet address of users). Because of this power, stewards are governed by their own policies: don't decide, don't promote users on projects with existing bureaucrats, don't change rights on your own project, act with transparency, and check local policies.[48] The "Don't Decide" policy further states:

> Stewards are not allowed to make decisions, such as "this user should (or should not) be promoted." Their task is to implement valid community decisions. . . . Stewards

should always be neutral. They can vote in elections, but when executing the result of the election the steward has to act according to the result, even if they disagree.[49]

At the time of incorporation in 2003, Wales delegated some of his authority to an initial five directors of the Wikimedia Foundation Board of Trustees, in which he serves as chairman emeritus. (The board has since been expanded; elections in July 2009 resulted in a total of ten trustees.[50]) The board "has the power to direct the activities of the foundation. It also has the authority to set membership dues, discipline and suspend members (article III), and to amend the corporate bylaws (article VI)."[51] In the realm of editorial disputes between users (including administrators) dispute resolution can be facilitated by mediation or arbitration, and the Arbitration Committee (ArbCom) can issue a binding decision. The ArbCom, discussed in chapter 5, was first proposed as a "Wikiquette committee" in 2003 and was formally established the following year.[52] However, it is recommended that disputes be worked out civilly between the participants as mediation and arbitration can be tedious. Or, as Skomorokh's Law notes, "There are no winners at Arbitration, only losers."[53] The ArbCom, the Board, and Jimmy Wales himself, ultimately, have the authority to penalize or remove abusive users.

Finally, while consensus is preferred for most decisions, voting has had a place in Wikipedia, such as in some elections (e.g., for stewards and board members) and on pages like "VfD" (Votes for Deletion) where allegedly unworthy articles are nominated for removal. Nonetheless voting is widely recognized as difficult and often contentious: "Don't vote on everything, and if you can help it, don't vote on anything."[54] In fact the VfD process was renamed to AfD (Articles for Deletion) and now speaks of consensus rather than voting.[55] In any case, and as noted earlier, multiple models of governance coexist within Wikipedia, and democratic voting is widely recognized as problematic.

However, despite an early lack of concern with community structure and culture (e.g., "Ignore All Rules"), protestations that administrators are nothing but janitors, and that the ArbCom was but an experimental delegation of authority from Wales, Wikipedia's conceptualization of governance and leadership is maturing and stabilizing. Wikipedia has long since recognized itself as a community, people strive to become administrators despite disclaimers, and the ArbCom is unlikely to go away. The cultural significance of administrators was acknowledged in January 2007 by the creation of the page "Advice for New Administrators," which became part of the "New

Admin School," which even includes the "coaching" (mentoring) of editors who want to become administrators.[56] Yet, the orthodox caveats about responsibility rather than power persist, as the "Advice" page cautions:

> Remember that administrator status is not a trophy. Generally, therefore, do not act any differently now than you did six months or a year ago. It is true that you may be able to help mediate a dispute effectively, or resolve one, or guide the improvement of an article. But in virtually all of these cases your ability has nothing to do with your being an administrator, just with your experience, knowledge of the policies, and good sense—i.e. virtues you had long before you became an administrator, and virtues shared by many non-administrators. . . . Wikipedia administrators do have certain powers, and you need good judgment to use them. Nevertheless, this does not mean that administrators should act like police or judges. Consider thinking of your new status more like a custodian.[57]

Furthermore, the role of socializing others into the collaborative norms of Wikipedia are represented as a central function of being an administrator, who should be willing to talk and be patient; respond with "gentle" encouragement and discouragement; pay "careful attention to our core policies"; "assume people act on good faith"; and "give people the benefit of the doubt." They should not "get sucked in" to the disputes in which they intervene.

Discussing Leadership

The prominent leader of an open content community is sometimes characterized as a benevolent dictator. This is a variation on a tradition in online communities, particularly Usenet, of referring to a minority with disproportionate influence as a "cabal."[58] While a cabal can still be spoken of in earnest (with a negative connotation), in time it and the acronym TINC ("There Is No Cabal") became shorthand for referring to the difficulties of community governance and the propensity for some to see conspiracies. The role of the "benevolent dictator" completes this ironic turn while also indicating genuine respect. Jimmy Wales is referred to as a benevolent dictator, though it is not a title he accepts. Indeed, it behooves any such leader to disclaim such a title because, as Eric Raymond notes, hacker culture "consciously distrusts and despises egotism and ego-based motivations; self-promotion tends to be mercilessly criticized, even when the community might appear to have something to gain from it. So much so, in fact, that the culture's 'big men' and tribal elders are required to talk softly and humorously deprecate themselves at every turn in order to maintain their status."[59] (Although Raymond is seminal for theorizing aspects of open source leadership and popularizing the term *benevolent dictator*, its usage appears to precede Raymond's use in computer communities and even its application to Linus Torvalds.[60])

Nonetheless, the need for "dictatorship" arises from the difficulty inherent to decision making in large, voluntary, and consensus-oriented communities. While a cabal or dictator might be complained about, so might their absence. In a discussion about whether a redesign of Wikipedia's portal should use icons of national flags to represent different languages—many nations share a language or use more than one—Wikipedian NSK wrote that continued arguments "do nothing to improve the present ugly portal." Unfortunately, "Wikipedia suffers from many voices, often contradictory. I think you need an influential leader to take final decisions (after community input of course)."[61] This sentiment is shared in many open content communities. FOSS practitioner Karl Fogel writes: "Only when it is clear that no consensus can be reached, and that most of the group wants someone to guide the decision so that development can move on, do they put their foot down and say 'This is the way it's going to be.'"[62] Clay Shirky also makes this point in his essay "A Group Is Its Own Worst Enemy" by way of Geoff Cohen's observation that "the likelihood that any unmoderated group will eventually get into a flame-war about whether or not to have a moderator approaches one as time increases."[63] (Again, Cohen's observation takes the form of the ever popular Godwin's Law.)

In the Wikipedia context, in addition to differing opinions among those of good faith, an informal and consensus-based approach does not seemingly deal well with those who act in bad faith, such as the feared neo-Nazi attack:

> What is needed in obvious cases like this is a "benevolent dictator," whether it's Jimbo Wales or the arbcom, to examine the editors' contributions then ban them, because these are not bona fide Wikipedians who happen to have a strong POV. They are fanatics acting to promote the views of a political cult, and they're here for no other reason. Yet here they remain, making a mockery of everything Wikipedia stands for.[64]

Where possible, Wales has delegated authority, particularly to the Board of Trustees and Arbitration Committee, but much authority remains with Wales as noted in a 2005 comment:

Wikipedia is "at the mercy of" Jimbo. Jimbo has delegated his "mercy," to use your term, to the Arbitration Committee that he convened over 15 months ago, and which he periodically refreshes the membership thereof as guided by the wishes of the community. Significant disciplinary matters in Wikipedia are thus guided by a number of editors who are held in high esteem by the community at large (or, at least, so one hopes).[65]

Anthere, a former chairperson of the board of trustees, described this balance of reserved authority and delegation as one of facilitating or hindering a direction, reminiscent of the goal theory of leadership[66] whereby a leader makes the subordinate's path more satisfying and easier to travel by clarifying goals and reducing obstructions:

I think that what is especially empowering is the leadership type of Jimbo. Jimbo is not coaching at all, and rather little directing (though hints are sometimes quite clear), as well as rather little delegating (I think the foundation would sometimes benefit from more delegation from Jimbo). His type is essentially supportive. Very low direction but very high support. This leaves basically as much opportunity to work in certain directions as one would dream of. However, one moves in a direction supported by Jimbo much more quickly than in a direction not supported by Jimbo. I[t] can take a long time to find a satisfactory decision, but prevents from travelling in an unsafe direction.[67]

However, this balance can lead to ambiguities that prompt discussion, such as that about editorial authority. In February 2005 an enormous debate erupted over the illustration included in the encyclopedic article on autofellatio. Images tend to prompt many debates and raise questions of censorship, free speech, cultural differences, and of the age appropriateness and quality of Wikipedia. A similar debate arose for the image in the clitoris article, as well as a cinematic still of Kate Winslet wearing nothing but a diamond necklace in the "Titanic (1997 film)" article. The latter debate was resolved when her breasts were cropped from the image;[68] it was eventually removed altogether because of copyright concerns. When Wales deleted the photographic image of autofellatio, which had replaced the less-contentious illustration, Erik Moeller challenged this action as it raised the old issue of to what extent Wikipedia has an "editor in chief":

Perhaps you could clarify that this was not done in your role as trustee. I don't believe it was, as you did not consult with Angela and Anthere [two other trustees], so I consider it just like an edit by any other Wikipedia editor, only that, of course, you hope that people will take it more seriously because of the reputation that comes with your role in the project, past and present. That's completely reasonable, if done rarely and in cases you consider important.

The page is currently being edit warred over, and one editor uses the comment "rv [revert] to Jimbo's approved version." It would be helpful if you could state here that you are not in the business of approving articles. I believe your edit summary "This image is completely unacceptable for Wikipedia" could be misconstrued to be an official statement, when it is your personal opinion. Some people still see Wikimedia as being governed by a benevolent dictator, and any explanation would help to eliminate that misconception.

I still remember how the Spanish Wikipedia forked over some discussion on advertising. I'm somewhat worried that people might misunderstand your comments, and assume that you are acting as "Chief Editor." On the other side, those who do support the removal of the image might deliberately seek to create that impression in order to further their agenda.[69]

Wales did not respond to this particular email message, but continued discussion with respect to how this image would affect educational use of Wikipedia. However, Wales's role was further discussed during discussion of the possible neo-Nazi attack. This led Wales to clarify that he would prevent such an attack though he also recognizes the dangers inherent to such action:

The danger of course is that the benign dictator may turn out to be biased or wrong himself. So I hesitate to do this except in cases where speed is of essence, or where it's just very clearcut and easy. What I prefer is that I can act as a temporary bridge and "person to blame" while we work on community solutions.[70]

Seven months later, on the same thread, Wales further defined his role as a "constitutional monarch":

I do not believe in the "benevolent dictator" model for Wikipedia. Our project is of major historical significance, and it is not appropriate for any one person to be the benevolent dictator of all human knowledge. Obviously.

But we have retained a "constitutional monarchy" in our system and the main reason for it is to _support_ and _make possible_ a very open system in which policy is set organically by the community and democratic processes and institutions emerge over a long period of experimentation and consensus-building. . . . It is not possible for 10,000 NeoNazis (if such numbers exist) to storm into Wikipedia and take it over by subverting our organic democratic processes because I will not allow it. Period. So we don't have to overdesign those processes out of a paranoia of a hostile takeover.

But this also means that we don't need to over-react right now. We can wait and see. They'll talk a big game but just review those message boards and then look around here. A battle of wits between Wikipedians and Nazis? I know who I'm betting on.[71]

Wales's conception of his role was further developed and articulated on the "Benevolent Dictator" discussion page:

I am more comfortable with the analogy to the British monarch, i.e. my power should be (and is) limited, and should fade over time. . . .

The situation in nl.wikipedia.org is probably a good example of how I can play a productive role through the judicious exercise of power. My role there is mostly just as advisor to people in terms of just trying to help people think about the bigger picture and how we can find the best ways to interact and get along to get our incredibly important work done.

But it is also a role of "constitutional" importance, in the sense that everyone who is party to the discussion can feel comfortable that whatever agreements are reached will be *binding*, that there is a higher enforcement mechanism. It's not up to me to *impose* a solution, nor is it up to me directly to *enforce* a solution chosen by the community, but I do play a role in guaranteeing with my personal promise that valid solutions decided by the community in a reasonable fashion will be enforced by someone. . . .

And notice, too, that I believe such authority should be replaced as time goes along by institutions within the community, such as for example the ArbCom in en.wikipedia.org, or by community votes in de.wikipedia.org, etc.

We have very few problems, other than isolated things, with sysop abuse or cabals, even in smaller languages, and in part because everyone is quite aware that I would take whatever actions necessary to ensure due process in all parts of wikipedia, to the best of my ability.[72]

It is worthwhile noting that Wales is articulating a hybrid of leadership types including autocratic (decision made by the leader alone), consultative (the problem is shared with and information collected from the group, before the leader decides alone), and delegated leadership (the problem is shared, ideas are accepted, and the leader accepts the solution supported by the group).[73] Also, Wales's concern with not over-designing the "organic democratic processes" echoes findings in the study of FOSS community that the judicious use of charismatic authority can be preferable to a "complex system of rules."[74] Similarly, in a discussion about the openness of foundation-related committees Wales felt that "it seems to me that the best way to approach this is not with a formalistic board resolution (this is not our traditional way), but through ongoing dialog and discussion, rather than rules-based demands from the board."[75] And even though Wales is seemingly conscientious about the use of his authority, others note that the "charismatic" character of his leadership can be unsavory. If others appropriate what Wales has said or done as the justification for their own position, some will object:

> This kind of hero-worship begins with Christians who find it more chic to parrot Christ's words than to live them. In our context this translates into using "Jimbo said . . ." as an argument that would stop all debate.[76]

Wales himself is now sensitive to this concern as seen in his qualification of an email about how to distinguish between sites that criticize Wikipedia and those that harass Wikipedians:

> I have this funny feeling, after writing this email, that it is the sort of email likely to be misused in some fashion as a WP:JIMBOSAYS fallacy. This note at the top serves as notice that anyone citing this email as setting down policy on Wikipedia is being a goof. I am just discussing and thinking here and trying to be helpful.[77]

Elsewhere he notes that "unless I am very very very careful, it ends up getting used as a stick to beat innocents to death with.:)"[78]

Concern about this role and title led to a consideration of alternatives for "benevolent dictator" including constitutional monarch, the most trusted party (TMTP, Linus Torvalds's preferred moniker), eminence grise, and deus ex machina.[79] And while the notion of constitutional monarch has achieved some stabilization and acceptance within the community, "benevolent dictator" won't disappear from the conversation given its long history within online communities. Indeed, the notion not only serves as a measure of the leader's actions, but also those of other participants. In one of the many threads about sexual content on Wikipedia, one participant wrote to another: "So your opinion is now law? Wonderful. We don't need all of those nasty little polls or votes. . . . All we have to do is have you make the decision for us. I thought Jimbo was the benevolent dictator. You seem just to want to be dictator, period."[80]

Conclusion

To whatever extent Wikipedia has been successful in the pursuit of a universal encyclopedia—a question for the next chapter—I argue an appreciation of the community and its collaborative culture is key to understanding Wikipedia. However, unlike the purity of a utopian dream, Wikipedians must reconcile their vision with the inescapable social reality of irritating personalities, philosophical differences, and external threats. Despite its good-faith collaborative culture, its egalitarian ethos, and its openness—or because of it—Wikipedia has been shaped by authorial leadership. An informal benevolent dictator serves to gently guide the community, to mediate internal disputes between those of good faith, and to defend against those acting in bad faith. At this point, he or she may achieve a significant amount of symbolic status within the community or even outside attention. However, when a person comes to be responsible for more than he or she can do by dint of will alone, new responsibilities and authority pull taut a tightrope that must be carefully walked before the eyes of one's peers. Sanger's reflections about his exit from the community and continued discussion about Wales's role are testaments to how delicately the tin crown of such leadership must be balanced.

7 Encyclopedic Anxiety

Over time, the average quality of Wikipedia articles rises, but Wikipedians' standards rise more quickly.
—OpenToppedBus's First Law

The higher the standards that Wikipedia aims for, the more that Wikipedia will appear sub-standard to the outside world.
—OpenToppedBus's Corollary

Wikipedia, and the collaborative way in which it is produced, is at the center of a heated debate. Much as reference works might inspire passionate dedication in their contributors, they also, seemingly, can inspire passionate disparagement. In 2004 Michael Gorman, former president of the American Library Association, wrote an op-ed criticizing Google and its book-scanning project; he was surprised by the negative online response to his piece, but this only prompted him to redouble his attack a few years later. In 2007 he focused on blogs and Wikipedia, decrying the effects of the "digital tsunami" on learning. In a blog essay entitled "Jabberwiki" Gorman lauded Sanger's abandonment of Wikipedia for the more expert-friendly Citizendium and criticized those who continue to contribute to, or even use, Wikipedia:

Despite Sanger's apostasy from the central tenet of the Wikipedia faith and his establishment of a resource based on expertise, the remaining faithful continue to add to, and the intellectually lazy to use, the fundamentally flawed resource, much to the chagrin of many professors and schoolteachers. Many professors have forbidden its use in papers. Even most of the terminally trendy plead with their students to use other resources. . . . A few endorse Wikipedia heartily. This mystifies me. Education is not a matter of popularity or of convenience—it is a matter of learning, of knowledge gained the hard way, and of respect for the human record. A professor who encourages the use of Wikipedia is the intellectual equivalent of a dietician who recommends a steady diet of Big Macs with everything.[1]

While he may be more strident than others, Gorman is not alone. As seen in the epigraph that begins this chapter, Wikipedians themselves are aware of the vertigo resulting from increasing quality being outpaced by expectations. What was once thought to be an adequate article, even when expanded and improved, might be marked as a stub today. As noted in the "corollary," this vertigo is further exaggerated in the "outside world's" view of Wikipedia progress.[2]

In this chapter I review some of the criticism Wikipedia faces related to the themes of collaborative practice, universal vision, encyclopedic impulse, and technological inspiration. However, I frame contemporary criticism by way of a historical argument: Wikipedia, like other reference works before it, has triggered larger social anxieties about technological and social change. This prompts the question of whether Wikipedia is representative of new forms of content production that are changing the role of the individual, the character of cultural products, and the authority and viability of established cultural institutions. Each element of this concern also prompts arguments about whether such changes are genuine or hype and, if genuine, positive or negative. But before I engage these specific arguments, it is worthwhile to understand why reference works prompt such arguments.

The Normativeness of the Reference Work

Many reference controversies revolve around the extent to which reference works are seen as *normative,* that is, in some way condoning their subjects and sources. For example, shouldn't a national dictionary shun popular slang or words borrowed from other languages?

When the French Academy commenced compiling a national dictionary in the seventeenth century, it was with the sense that the language had reached perfection and should therefore be authoritatively "fixed," as if set in stone. However, the utilitarian value of a vernacular dictionary could not be denied: Furetiére's competing dictionary contained words not approved of by the scholars and it sold well in the black market.[3] Samuel Johnson also thought he might be able to preserve the purity of English, despite warnings that the French dictionary took forty years to complete.[4] However, once the difficult task of compiling a dictionary was complete, he apologized in its preface to those "who have been persuaded to think well of my design, [and] require that it should fix our language"; this pretense with which he had "flattered" himself was in fact an "expectation which neither reason nor experience could justify" as no lexicographer can secure his language "from corruption and decay" and "clear the world at once from folly, vanity, and affectation."[5]

While there is still some debate over the extent to which dictionaries should be "prescriptive,"[6] few, beside the French Academy, would purposely exclude commonly used words out of a desire to withhold implicit approbation. However, encyclopedists have been more willing to associate the scope of their subject, and its treatment, with a larger social program. One reason for this difference between dictionaries and encyclopedias might simply be space. It is within the realm of a lexicographer to include every word of interest, even if it requires twenty volumes in the case of the *OED* (*Oxford English Dictionary*), or a magnifying lens in the case of the *OED*'s compact edition. Encyclopedias, if they are to fit on one or two shelves of a library stack, must limit their scope. This then requires judgment about what to include in a given work, which entails asking what is essential, worthwhile, and appropriate to know. On the axis of material constraints then, Wikipedia is situated much more like paper dictionaries than encyclopedias given its near-infinite number of pages. (Granted, Wikipedians still argue about inclusionism versus deletionism,[7] but even a deletionist's scope is far more permissive than even the largest print encyclopedia.)

Another probable reason reference works are thought to be normative is that they were marketed as resources for children. Information historian Foster Stockwell concludes,"The implication was that any parent who failed to buy an encyclopedia for the youngster was depriving a child of the opportunity of doing well in school, and, ultimately, in life." Between 1940 and 1970 some sales techniques were so aggressive as to be outlawed and various encyclopedias were fined for violating Federal Trade Commission orders. Yet, despite the scholarly intentions of their compilers, the marketing departments of reference work publishers convincingly made their pitch and the public came to see encyclopedias as an authoritative source for instruction, such that, "when children go to their parents for help they will, as often as not, be directed to the encyclopedia shelf."[8] This issue is reflected today in arguments about Wikipedia's age appropriateness: is it "child safe"? The English-language Wikipedia has generally resisted content discrimination on the basis of anything other than informative content, though how to deal with potentially offensive subjects is often discussed (e.g., pedophilia and hate speech). The Wikimedia Foundation addresses concerns about age appropriateness partly through the provision of a Simple English Wikipedia for use by children.[9]

Other wiki-based projects face a similar issue. The very handy wikiHow provides accessible information on how to do various tasks yourself; yet, just because a page describes how to do something, does that mean one should do it? (An article on wikiHow about *how to do* something compared to an article on Wikipedia *about* something seems to have a greater force.) wikiHow makes no claim that every article is an endorsement, but it also avoids content that would be considered "inappropriate for our family audience,"[10] a threshold the larger Wikipedia does not accommodate. Despite these intentions, and any disclaimers, some people nonetheless see Wikipedia as representative and permissive of changes not to their liking. In the history of reference work production, Wikipedia is not alone.

This question of an implied morality in a reference work is present in Herbert Morton's fascinating *The Story of Webster's Third: Philip Gove's Controversial Dictionary and Its Critics*.[11] Perhaps the primary reason for the controversy associated with this dictionary was that it appeared at a time of social tumult. A simplistic rendering of the 1960s was that progressives were seeking to shake up that which conservatives held dear. Yet, those working on the *Third* were not a band of revolutionaries. For example, Gove made a number of editorial decisions so as to improve the dictionary. And while lexicographers might professionally differ with some of his choices, such as the difficult pronunciation guide or the sometimes awkward technique of writing the definition as a single sentence, these were lexicographic decisions. It was the social context that largely defined the tenor of the controversy.

My reading of Morton, and one I think is relevant to Wikipedia as well, is that critics were alarmed at the social change occurring around them and attacked *Webster's Third* as an exemplar and proxy. For example, Wilson Follet, an authority on word usage, published an article in the *Atlantic Monthly* entitled "Sabotage in Springfield" wherein he described the *Third* as "a scandal and disaster," which "plumes itself on its faults and parades assiduously cultivated sins as virtues."[12] Scholar Jacques Barzun thought it extraordinary, and worth bragging about, that for the first time in his experience the editorial board of the distinguished *American Scholar* was able to unanimously condemn a work and know where each board member "stood on the issue that the work presented to the public," even though "none of those present had given the new dictionary more than a casual glance."[13] In fact, an exhortation I encountered as a schoolboy of "ain't ain't a word" was a prominent topic of national debate after the *Third*'s publication.[14] Yet, as Morton details, while some of these criticisms resulted from Merriam's ill-considered press materials proclaiming it to be "truth," "unquestionable fact," and the "supreme authority," much of the reaction was also predicated on ignorance and a reaction against "the so-called permissiveness of American culture in the 1960s."[15] The extent to which Wikipedia makes claims of veracity or greatness is part of the debate I will discuss.

Bias: Progressive and Conservative

In 2008 the front page of Conservapedia, an ideological competitor of Wikipedia, recommended its "article of the year" to readers so they might "Discover what Wikipedia, the public school systems, and the liberal media don't want you to know about atheism."[16] This reference to the "Atheism" article clearly indicates Conservapedia's intention of opposing a perceived liberal and materialistic bias in Wikipedia. Indeed, its "Examples of Bias in Wikipedia" article lists 160 instances.[17] And Conservapedia is but one of the first of many ideological user-generated encyclopedias likely to be started—though many soon fall into disuse. (The facetious headline of an article in the *Register* recommends that if you find "Conservapedia too pinko? Try Metapedia."[18] Metapedia's stated purpose is to serve as an encyclopedia "for pro-European activists,"[19] recalling the much discussed "neo-Nazi" attack/fork of Wikipedia.) Because reference works are popular, used by children, and understood as representing what is known, we should not be surprised to see these works at the center of larger social controversies.

And because of visionaries like Otlet and Wells one might mistakenly infer that reference works are necessarily progressive. While this has often been the case, particularly since the Enlightenment, it need not be so. In the history of reference works one is more likely to find opposing forces, cycles of predominance, and surprises. As an example of the diversity of purpose for reference works, historian Tom McArthur claims the Greeks wanted to know everything so as to think better, the Romans to act better, and the Christians to glorify God and redeem their sins.[20] As evidence of the latter Johann Zedler wrote in his eighteenth-century encyclopedia, the *Universal-Lexicon*: "the purpose of the study of science . . . is nothing more nor less than to combat atheism, and to prove the divine nature of things."[21] In Conservapedia's "Atheism" article of the year, we see the cycle has completed a turn.

Of course, it is the French *Encyclopédie* with which progressivism is famously associated. In its "Encyclopedia" article, Diderot wrote that a good encyclopedia ought to have "the power to change men's common way of thinking."[22] Such a notion was considered dangerous by the French nobility, Pope Clement XIII, and an editor of *Britannica*, a clergyman by the name of George Gleig. In the dedication of the 1800 *Britannica* supplement, Gleig wrote to his monarch: "The French *Encyclopédie* has been accused, and justly accused, of having disseminated far and wide the seeds of anarchy and atheism. If the *Encyclopædia Britannica* shall in any degree counteract the tendency of that pestiferous work, even these two volumes will not be wholly unworthy of Your Majesty's attention."[23] Wikipedia is often thought to be anarchic as well, or at least to be an experiment in anarchism—and the recurrent motif of concern about atheism is remarkable. However, ironically, *Britannica*'s image as a conservative stalwart is contradicted by one of its more recent editors, Charles Van Doren; Jimmy Wales is fond of citing the former editor at *Britannica* as saying that "because the world is radically new, the ideal encyclopedia should be radical, too. It should stop being safe—in politics, and philosophy, and science."[24] The fact that Van Doren worked at *Britannica* after resigning from Columbia University because of his participation in the television quiz show scandals of the 1950s is a further irony given the present arguments about new media and the authority of knowledge production.[25]

Accusations of bias are surprising in their specificity and passion, and prior to Wikipedia, *Britannica* received the brunt of attention. Herman Kogan's *The Great EB: The Story of the Encyclopædia Britannica* addresses many accusations of bias, particularly by and between Protestants, Catholics, Britons, Americans, and Soviets.[26] Harvey Einbinder's *The Myth of the Britannica* is actually an extensive criticism itself though he also describes Christian Scientist and Jehovah's Witness concerns in addition to Catholic controversies.[27] Over a century ago Thaddeus Oglesby collected criticisms he had raised against *Britannica* in a book entitled: *Some Truths of History: A Vindication of the South against the Encyclopædia Britannica and Other Maligners*.[28] (However, contrary to Oglesby's opinion, Gillian Thomas notes that the *Britannica* seems overly favorable to the South given its portrayal of lynching as a form of controlling "disorderly Negro politicians" by "protective societies of whites."[29]) More recently, Michel McCarthy wrote of *Britannica*'s complaint department including their receipt of an obscenity-filled letter from a Texas man accusing *Britannica* of bias against the Ostrogoths.[30]

But perhaps the best-known encyclopedia critic is Joseph McCabe; around 1950 he began documenting a perceived Catholic bias in many popular encyclopedias. McCabe's dedication and focus has the same obsessive character of earlier reference work compilers, present-day Wikipedians, and even some of its critics. He wrote that this new preoccupation resulted from an overseas argument about the Pope's employment of castrati. He discovered that *Britannica's* once accurate "Eunuchs" article "had been scandalously mutilated, the facts about church choirs suppressed, and the reader given an entirely false impression." Upon learning that the Westminster Catholic Federation boasted of their efforts to "eliminate matter which was objectionable from a Catholic point of view and to insert what was accurate and unbiased," McCabe set out to identify what had been altered in his publication *The Lies and Fallacies of the Encyclopædia Britannica: How Powerful and Shameless Clerical Forces Castrated a Famous Work of Reference*.[31] He followed this work a few years later with *The Columbia Encyclopedia's Crimes against the Truth: Another Analysis of Potential Catholic Bias in Encyclopedia*. Here he tracked changes in various editions over the topics of sexuality, atheists, the forgery of the *Donation of Constantine* (transferring power from the Roman Emperor to the papacy), and the encyclopedia's silence on "Catholic persecution, death sentence for heresy, mental reservation, apostates, vilification marriages, torture, Feast of Fools, the Syllabus, etc."[32] No doubt, he would have loved to have a tool like WikiScanner.[33] This tool, which can help identify the origins of some "anonymous" edits, was widely covered in the press in August 2007 when it was revealed that computers associated with the Diebold electronic voting machine company, the Democratic Congressional Campaign Committee, the Vatican, Scientology members, and others had removed embarrassing information from their respective articles.

Yet, as in any history, we must be careful not to divide the field into extremes, in this case between conservative and progressive poles. For example, while an association with the *Encyclopédie* was certainly dangerous, Robert Darnton notes that it was France's sympathetic director of the library, and chief censor, who saved the *Encyclopédie* several times. Indeed, Malesherbes warned Diderot that his papers were about to be seized by the police but that they could be deposited and saved with him after issuing the very order for their confiscation.[34] Or, in another anecdote, one can see that even the French Royals had a complicated relationship with the censored work, wishing they had the reference on hand during a dinner party discussion about the composition of gunpowder and the construction of silk stockings.[35]

And as a final methodological note, the interpretation of past events is often colored by our own present. Consider the question, what did those in power fear from the *Encyclopédie*? Foster Stockwell clearly labels the focus on craftsmanship as a progressive force: Diderot "exploded the religious and social myths that kept people in a condition of servitude." He was also the first to take craftsmanship seriously and by doing so "helped set in motion the downfall of the royal family and the rigid class system," with the result that "every person became the equal of every other."[36] Yet another scholar, Cynthia Koepp, renders the import quite differently. Diderot, on behalf of the "dominant, elite culture" expropriated the techniques of the artisan whose "formally unique talents, knowledge, and abilities became dispensable."[37]

The difference between these two authors shows that the degree to which reference works are viewed as conservative or progressive is not only dependent on their historical context, but also on interpretations of that history in the present: Stockwell sees the *Encyclopédie* as a democratizing force whereas Koepp sees it as a form of expropriation. (It could have been neither or both.) Consequently, the task is not so much to determine whether a particular reference work was objectively and definitively conservative or progressive, but rather whether it was received as such and what that tells us of the larger social context. As Einbinder writes in the introduction to his critique, "since an encyclopedia is a mirror of contemporary learning, it offers a valuable opportunity to examine prevailing attitudes and beliefs in a variety of fields."[38] Similarly, for contemporary debate, Clay Shirky, a theorist of social software, observes: "Arguments about whether new forms of sharing or collaboration are, on balance, good or bad reveal more about the speaker than the subject."[39]

Criticisms of Wikipedia and "Web 2.0"

Not surprisingly, though worth a chuckle nonetheless, an informative resource for this chapter is Wikipedia's "Criticism of Wikipedia" article. It contains the following dozen or so criticisms of the the Wikipedia concept and its contributors:

> *Criticism of the concept*: the wiki model, usefulness as a reference, . . ., suitability as an encyclopedia, anti-elitism as a weakness, systemic bias in coverage, systemic bias in perspective, difficulty of fact-checking, use of dubious sources, exposure to vandals, exposure to political operatives and advocates, prediction of failure, privacy concerns, quality concerns, threat to traditional publishers, "waffling" prose and "antiquarianism," anonymous editing, copyright issues, the "hive mind." *Criticism of the contributors*: flame wars, fanatics and special interests, censorship, abuse of power, level of debate, male domination, community, EssJay and the lack of credential verification, humorous criticism.[40]

Those are substantive concerns raised about Wikipedia—each interesting in its own way—and many are responded to on another page.[41] Also, many of the specific complaints are part of a more general criticism in which Wikipedia is posed as representative of an alleged "2.0" shift toward a hivelike "Maoist" collective intelligence. The term *Web 2.0*, unavoidable in a discussion about Wikipedia, is attributed to a conversation about the naming of a conference in 2004 to discuss the reemergence of online commerce after the collapse of the 1990s "Internet bubble." Tim O'Reilly, technology publisher, writes that chief among Web 2.0's "rules for success" is to: "Build applications that harness network effects to get better the more people use them. (This is what I've elsewhere called 'harnessing collective intelligence.')."[42] However, many of the platforms claimed for Web 2.0 preceded it, including Amazon, Google, and Wikipedia. Ward Cunningham launched his wiki in 1995! I'm forced to agree with Robert McHenry, former editor in chief of *Encyclopædia Britannica*, that "Web 2.0" is a marketing term and shorthand "for complexes of ideas, feelings, events, and memories" that can mislead us, much like the term "the 60s."[43] (The label *modern* can be equally frustrating, as we shall see.)

Fortunately, while unavoidable, one can substantiate the notion of Web 2.0 by focusing on *user-generated content.* Clay Shirky, in *Here Comes Everybody,* argues we are moving from a model of "filter then publish" toward "publish then filter"; filtering before was by publishers, today it is by one's peers.[44] This seems to be the most important feature of "2.0," one represented by Craigslist postings, Amazon book reviews, blog entries, and Wikipedia articles. The production of content by Shirky's "everybody" or Wikipedia's "anyone" is what Wikipedia's collaborative culture facilitates and what its critics lament.

In the following sections I engage criticism of Wikipedia, and Web 2.0 more generally, via four themes: collaborative practice, universal vision, encyclopedic impulse, and technological inspiration. In short, a caricature of the criticism that I address is that the fanatical mob producing Wikipedia exhibits little wisdom and is more like a Maoist cult of monkeys banging away on the keyboards and thumb pads of their gadgets, disturbing the noble repose of scholars and displacing high-quality content from the marketplace. Though I am personally sympathetic toward Wikipedia, my intention is not to argue for or against Wikipedia supporters or critics, but to identify the larger social issues associated with Wikipedia collaboration and the pursuit of the universal encyclopedia.

Collaborative Practice

In many conflicts misunderstandings are as common, if not more so, than genuine differences. There are elements to this in the arguments about Wikipedia, particularly over the way it is produced. Describing how knowledge is constituted can be difficult, but one can identify three ways for how we might think of knowledge production throughout history.[45] First, we must admit that the hermit's encyclopedia, devoid of all contact with the words of others, would be of little use. Even the monastic scribe copying parchment, and introducing some changes no doubt, is engaged in some degree of sociality. Some have described this interaction at a distance, in time or geography, as a type of stigmergy, like a wasp building upon the work of others.[46] Second, the production of a reference work eventually exceeded the capability of any one person. What might be thought of as corporate production includes the interaction of financiers and subscribers, and of contributors and editors working within some—even if loose—form of social organization. Finally, there is Wikipedia and other open content.

In earlier chapters I explore how this community and its culture facilitate the production of an encyclopedia. It is on this point that there is much argument, and, I think, some misunderstanding. The central concern seems to be how we can conceive of our humanity in working together, and its implications. (If this sounds confusing or overly grand, bear with me!) I'll begin with two related buzzwords: the *hive-mind* and *collective intelligence*.

A hot topic of the 1990s was chaos and complexity theory; Kevin Kelly, former editor in chief of *Wired*, published a popular book on the topic entitled *Out of Control: The New Biology of Machines, Social Systems, and the Economic World*.[47] Kelly popularized a burgeoning understanding of how order can emerge from seeming chaos: how the beautiful midair choreography of a flock of birds arises when many individuals follow very simple rules of interaction. This "new biology" was mostly gleaned from and applied to the natural world, but Kelly also posited it as a theory in understanding social organization and intelligence via the notion of the "hive mind." This idea would persist into the new millennium when a number of new media-related phenomena arose demanding explanation. In 2002 Howard Rheingold, who had previously authored a seminal and popular treatment of virtual communities, published *Smart Mobs: The Next Social Revolution*.[48] In this book Rheingold argues for new forms of emergent social interaction resulting from mobile telephones, pervasive computing, location-based services, and wearable computers. Two years later, in *The Wisdom of Crowds*, James Surowiecki made a similar argument, but instead of focusing on the particular novelty of technological trends, he engaged more directly with the social science of group behavior and decision making.[49] In his book Surowiecki argues that groups of people can make very good decisions when there is diversity, independence, decentralization, and appropriate aggregation within the group. This works well for problems of cognition (for which there is a single answer) and coordination (by which an optimal group solution arises from individual self-interest, but requires feedback), but less so for cooperation (for which an optimal group solution requires trust and group orientation, i.e., social structure/culture).

None of these authors engage the case of Wikipedia, which was just beginning to receive significant press coverage in the latter half 2001. But since the publication of *Smart Mobs* and *The Wisdom of Crowds*, two questions have arisen: Are these works on group dynamics applicable to understanding Wikipedia's apparent success; and if so, is that a good thing? But let's begin with the latter question first: many Wikipedia critics think the collective intelligence model is applicable, and are repulsed by the process and the result.

Michael Gorman, the acerbic librarian encountered at the beginning of this chapter, writes: "The central idea behind Wikipedia is that it is an important part of an emerging mass movement aimed at the 'democratization of knowledge'—an egalitarian cyberworld in which all voices are heard and all opinions are welcomed."[50] However, the underlying "'wisdom of the crowds' and 'hive mind' mentality is a direct assault on the tradition of individualism in scholarship that has been paramount in Western societies." Furthermore, whereas this enthusiasm may be nothing more than easily dismissible "technophiliac rambling," "there is something very troubling about the bleak, dehumanizing vision it embodies—this monster brought forth by the sleep of reason."[51] In a widely read and discussed essay entitled "Digital Maoism: The Hazards of the New Online Collectivism," Jaron Lanier, computer scientist and author, concedes that decentralized production can be effective at a few limited tasks, but that we must also police mediocre and malicious contributions. Furthermore, the greatest problem is that the hive mind leads to a loss of individuality and uniqueness: "The beauty of the Internet is that it connects people. The value is in the other people. If we start to believe the Internet itself is an entity that has something to say, we're devaluing those people and making ourselves into idiots."[52] Andrew Keen, 1990s Internet entrepreneur turned Web 2.0 contrarian, likens the process of producing and consuming content to "the blind leading the blind—infinite monkeys providing infinite information for its readers, perpetuating the cycle of misinformation and ignorance."[53] Author Mark Helprin, like Gorman, unwittingly stepped on a hornets' nest of online dissent with an op-ed. His call to extend copyrights in the United States prompted a backlash that he responded to with a book defending his proposal and counter-attacking the "functionally illiterate" digital barbarians. Free-culture advocates protest the moving window of copyright extensions because, they argue, it creates a perpetual copyright; when copyright terms are continually extended by Congress this contravenes the intentions and limits specified in the U.S. Constitution. Because he was apparently unaware of this controversy and shocked by the vociferous response, he likened the way people work together online as termites that "go steadily and quietly about their business, almost unnoticed" until "an apparently solid house collapses in a foamy heap."[54]

(Lawrence Lessig's review characterizes Helprin's book as an odd combination of memoir and poorly informed policy.[55])

Yet the question of whether this model is actually relevant to Wikipedia is disputed by many, including prominent Wikipedians. In May 2005 Wikipedian Alex Krupp introduced Surowiecki to the wikipedia-l list via a message entitled "Wikipedia, Emergence, and The Wisdom of Crowds":

> I think all Wikipedians would enjoy the book. . . . The basic premise is that crowds of relatively ignorant individuals make better decisions than small groups of experts. I'm sure everyone here agrees with this as Wikipedia is run this way.[56]

Jimmy Wales was quick to respond that he did not agree, and stressed as much in his public talks because "Wikipedia functions a lot more like a traditional organization than most people realize—it's a community of thoughtful people who know each other, not a colony of ants."[57] Another Wikipedian expressed a similar sentiment based on his experience that Wikipedia is built by "dedicated editors collaborating and reasoning together . . . it is hard to recognize the effect, if any, of 'swarm intelligence' on the project's development."[58]

I participated in the thread myself, hoping to move beyond the *swarm* label toward why the theory might be relevant to Wikipedia production if it can be characterized by diversity, independence, and decentralization within the group. In particular, these conditions might augment other theorists' explanations of "commons-based peer production":

> If the asynchronous and bite-sized character of Open contributions contribute to their success (Benkler "fine-grained," Sproull "microcontributions"), is that all? What *kind* of micro-contributions are necessary? *If* the contributions are crap, if they aren't coming from diverse participants (e.g., not "group think"), independent (e.g., not "herding"), and decentralized and filtered/aggregated well (e.g., not "US intelligence" ;)) then they might be useful.[59]

However, even the premise of my point was disputed: what role did diverse, sometimes anonymous, fine-grained micro-contributions play in Wikipedia production? Scholars Yochai Benkler and Lee Sproull were among the first to argue the importance of such contributions in online communities.[60] However, while present, how relevant was this for Wikipedia production? Ward Cunningham has identified openness and incrementalism as key design principles of the wiki.[61] Others focused on the fact that a relatively tight-knit minority did the majority of the work and the majority did little, often explained by way of theories of the long tail, Pareto's

Distribution, Zipf's Law, or the 80/20 Rule; or that we were witnessing the power of "mass collaboration."[62] Oddly, as reviewed in chapter 1, two seemingly contrary popular theories were being used to explain Wikipedia at the same time: is the crowd or the elite doing a majority of the work?

In any case, the important point was that Wikipedians typically rejected any characterization of Wikipedia as some sort of smart mob, as Wales did:

> I should point out that I like Surweicki's thesis just fine, it's just that I'm not convinced that "swarm intelligence" is very helpful in understanding how Wikipedia works—in fact, it might be an impediment, because it leads us away from thinking about how the community interacts in a process of reasoned discourse.[63]

Of course, as is evident in my concern with Wikipedia culture in earlier chapters, I wrote that I agreed "with Jimbo that any posited explanation that fails to account for the dynamics and culture of good-willed interaction has got it wrong. So in that sense, Surowieki is (perhaps) necessary but (certainly) not sufficient."[64] Yet, despite an admittedly incomplete understanding and Wales's public attempts to disclaim Wikipedia as a "hive-mind," the accusation continues to be raised. For example, in September 2006 an otherwise informative article entitled "The Hive: Can Thousands of Wikipedians Be Wrong?" appeared in the *Atlantic Monthly*.[65] In addition to likening online collaboration to a hornet's nest and termite infestation, Helprin made an even less favorable comparison: these interactions were like a "quick sexual encounter at a bacchanal with someone whose name you never know and face you will not remember, if, indeed, you have actually seen it." The resulting works "are often so quick, careless, and primitive that they are analogous to spitting on the street."[66] In his 2006 "Digital Maoism" essay, Lanier recast the claim of the hive as implying inevitable incremental improvement: "A core belief of the wiki world is that whatever problems exist in the wiki will be incrementally corrected as the process unfolds."[67] The essay was also published with commentary from a number of prominent thinkers. Wales responded that Lanier's allegation was unfounded:

> . . . this alleged "core belief" is not one which is held by me, nor as far as I know, by any important or prominent Wikipedians. Nor do we have any particular faith in collectives or collectivism as a mode of writing. Authoring at Wikipedia, as everywhere, is done by individuals exercising the judgment of their own minds.[68]

Yochai Benkler, law professor and seminal theorist of "commons-based peer production" also responded: "Wikipedia is not faceless, by and large.

Its participants develop, mostly, persistent identities (even if not by real name) and communities around the definitions."[69] Addressing the question of collectivism and the implication of rosy utopianism, Clay Shirky noted, "Wikipedia is the product not of collectivism but of unending argumentation; the corpus grows not from harmonious thought but from constant scrutiny and emendation."[70]

Contrary to the allegations of critics, Wikipedia supporters argued that wikis were both a powerful tool "that fosters and empowers responsible individual expression"[71] and a community of peers working within a collaborative culture—neither of which was best described by the notion of a swarm, hive, or collective intelligence. Indeed, it seems that the actual understanding of Wikipedia supporters is not that different from Gorman's conception of an encyclopedia. Gorman claims that whereas a traditional encyclopedia is "the product of many minds," it is not "the product of a collective mind." Instead, "It is an assemblage of texts that have been written by people with credentials and expertise and that have been edited, verified, and supplied with a scholarly apparatus enabling the user to locate desired knowledge."[72]

The real issue to explore, then, is the extent to which access to encyclopedic production is provided to those without "credentials and expertise."

Universal Vision

A simple summary of the universal encyclopedic vision is its aspiration of expansiveness. Otlet's "Universal Repertory" and Wells's "World Brain" were conceived of as furthering an increased scope in production and access. Reference work compilers would be joined by world scholars and international technocrats. Furthermore, every student might have these extensive resources at hand, in a personal, inexpensive, and portable format. It was hoped this collection of intellect would yield a greater sense of mutual accord throughout the world. Nor would the world encyclopedia limit itself to text; new media and tools were accommodated and envisioned. The universal vision persisted into the networked age, becoming more modest in its hope of prompting world peace, but pushing accessibility even further. Once Project Gutenberg launched, content could be had for the cost of network access; then, as access became pervasive, information became free "as in beer"; and then, in Stallman's proposal for a "Universal Encyclopedia," content was to become free "as in freedom": free to be distributed and modified without restriction, other than reciprocity.[73] In the Interpedia days it was thought that reasonable and well-educated people might contribute—which was how most Internet users could conceive of themselves at the time. Nupedia, too, had the potential to open up contribution, even if it was still limited to the formally educated. And, of course, with Wikipedia almost "anyone" can edit, something not even conceived of—or perhaps even approved of—by the earliest visionaries.

Critics of Wikipedia find this to be a cockeyed dream that is quickly becoming an all-too-real nightmare, and liken the universal vision to failed utopias and feared dystopias. In a *Wired* profile of Tim O'Reilly, journalist Steven Levy touches upon the Internet and collective consciousness, and asks if Web 2.0 might be "the successor to the human potential movement"; Nick Carr, a journalist covering information technology, claims that even entertaining this question is evidence of unhinged rapturous "revelation."[74] Michael Gorman equates the Internet with the siren song that lures sailors to shipwreck.[75] Thomas Mann, another librarian, invokes Aldous Huxley in an essay entitled "Brave New (Digital) World"—subtitled "Foolishness 2.0"—and compares the vision of user-generated content to naïve French and Marxist revolutionaries. Mann argues we would be better served emulating the pragmatic authors of the *Federalist Papers*, cognizant of the pathologies that infect social organisms (e.g., that "short-sightedness, selfishness, and ignorance are constant factors in human life"), rather than celebrating the unproven presumption that technology can cure all.[76]

In this case, the larger anxiety that Wikipedia has triggered is clear, and like that of its predecessors it reflects a broader concern about authority. Much as the *Encyclopédie* challenged the authority of church and state and recognized the merit of the ordinary artisan, or the *Third* reflected larger social changes manifested in everyday speech, Wikipedia is said to favor mediocrity over expertise. Or from Andrew Keen's perspective, Wikipedia elevates *The Cult of the Amateur* at the expense of the professional.

The implication of this shift toward user-generated content and niche markets is contested. Or, it is not so much that different authors envision different futures, but viscerally react to that same future differently. (However, we should remember that all those characteristics now associated with print—its "fixity," authority, and credibility—cannot be taken for granted and their establishment took some time to develop as a "matter of convention and trust, of culture and practice."[77]) The popular InstaPundit blogger

Glenn Reynolds has a positive interpretation as seen in the title of his book: *An Army of Davids: How Markets and Technology Empower Ordinary People to Beat Big Media, Big Government, and Other Goliaths.*[78] And Chris Anderson, the current editor in chief of *Wired,* finds "selling less for more" in *The Long Tail* to be the exciting future of business because retailers can now offer easy access to the "long tail" of niche markets in which that majority of items only sell a few copies.[79] However, on the flip side, Keen argues that "today's Internet is killing our culture." Keen begins his book by mourning the closing of Tower Records, a favorite of his in which he could peruse, hands on, a deep and diverse catalog of music. Independent bookstores and small record labels have also disappeared, and should rampant piracy and the flood of mediocre user-generated content continue, other creative industries face the same fate. Yet, while Keen laments the effects of a cult, Anderson finds value in the long tail: celebrating the easy access and massive selection of Amazon (for books), Rhapsody (for music), and Netflix (for movies).

However, besides implications for the marketplace, the question of authority also invokes concerns about autonomy and liberty. Matthew Battles, a journalist and librarian, responds to critics who prefer the professional to the amateur by asking who is going to force the cat back in the bag:

> Does Gorman really believe, along with Andrew Keen, that "the most poorly educated and inarticulate among us" should not use the media to "express and realize themselves"? That they should keep quiet, learn their place, and bow to such bewigged and alienating confections as "authority" and "authenticity"? Authority, after all, flows ultimately from results, not from such hierophantic trappings as degrees, editorial mastheads, and neoclassical columns. And if the underprivileged (or under-titled) among us are supposed to keep quiet, who will enforce their silence—the government? Universities and foundations? Internet service providers and media conglomerates? Are these the authorities—or their avatars in the form of vetted, credentialed content—to whom it should be our privilege to defer?[80]

Shirky similarly notes the "scholars-eye view is the key to Gorman's complaint: so long as scholars are content with their culture, the inability of most people to enjoy similar access is not even a consideration."[81]

This concern about access and authority is further manifested by way of argument about two labels: modernism and Maoism. Matthew Battles, continuing his response on authority, argues that genuine "digital Maoism" emerges when users are bullied to be kept silent:

Experience, expertise, and authority do retain their power on the web. What's evolving now are tools to discover and amplify individual expertise wherever it may emerge. Maoist collectivism is bad—but remember that Maoism is a thing enabled and enforced by authority. Similarly, digital Maoism rears its head whenever we talk about limiting the right to individual expression that, with the power of the web behind it, is creating a culture of capricious beauty and quirky, surprising utility. Digital Maoism will emerge when users are cowed by authority, when they revert to the status of mere consumer, when the ISPs and the media conglomerates reduce the web to a giant cable TV box.[82]

Interestingly, critics and supporter alike recognize threads of Enlightenment and modern values in contemporary knowledge work. In their own way, supporters and critics each lay claim. In June 2007 *Encyclopædia Britannica* hosted an extensive "Web 2.0 Forum" on its blog, upon which danah boyd, a researcher of online communities and a PhD student at the time, declared:

I entered the academy because I believe in knowledge production and dissemination. I am a hopeless Marxist. I want to equal the playing field; I want to help people gain access to information in the hopes that they can create knowledge that is valuable for everyone. I have lost faith in traditional organizations leading the way to mass access and am thus always on the lookout for innovative models to produce and distribute knowledge.[83]

Two points are worthwhile noting about this comment. First, boyd—who prefers her name in lowercase—is comparing new knowledge production models with that of the traditional academy, something she implies some dissatisfaction with here and more pointedly elsewhere.[84] Historian Peter Burke argues that the institutions of the university, academy, and scholarly society each arose when its predecessor failed to accommodate new approaches to knowledge production and dissemination[85]—perhaps Wikipedia stands astride another such fault. Second, boyd self-identified—I assume sincerely—as a Marxist, and this merits some framing. A common insult levied against those in the free culture movement is the aspersion of communism—or socialism and now even Maoism.[86] Such statements are usually received as an insult, as intended, and denied. Indeed, given the strong libertarian roots of Internet culture it is a grave mistake to accept such a generalization. Jimmy Wales, a former futures and options trader, credits Friedrich Hayek, a famous free market thinker, with informing his understanding of collective behavior.[87] In any case, despite red-baiting or parading, one should remember that Karl Marx was as "modern" as Adam

Smith; by this I mean that although their mechanisms of social action were different, each was relatively optimistic about the power of human beings to positively shape their own destiny.

The critics too, will admit to a modern streak: Mann writes that modernism was a good thing, but presently "people's faith in the transformative effects of gadgets" is utopian, and as Gorman points out, a siren song.[88] Gorman himself responds in an essay about Google's efforts to scan millions of books:

> How could I possibly be against access to the world's knowledge? Of course, like most sane people, I am not against it and, after more than 40 years of working in libraries, am rather for it. I have spent a lot of my long professional life working on aspects of the noble aim of Universal Bibliographic Control—a mechanism by which all the world's recorded knowledge would be known, and available, to the people of the world. My sin against bloggery is that I do not believe this particular project will give us anything that comes anywhere near access to the world's knowledge.[89]

Keen too, while critical of Wikipedia, refuses to cede the label of modern. In response to Wales describing himself in a widely read article as "very much an Enlightenment kind of guy," Keen argues that Wales "is a counter-enlightenment guy, a wide-eyed-dramatic, seducing us with the ideal of the noble amateur."[90] At this point, as is the case with "Web 2.0," I balk. I don't question that it is convenient to use a label commonly associated with a historical period so as to evoke a common understanding of the prominent events and related social themes. However, should we want to argue about whether something is, or is not, modern it is best if we ground that discussion with theoretical clarity and historical specificity. Otherwise, we may be speaking past each other—this is why I speak of a twentieth-century universal aspiration, encyclopedic impulse, technological inspiration, and collaborative practice.

In any case, in this argument about how Wikipedia is collaboratively produced we see a larger argument about authority, its institutions, individual autonomy, as well as possible consequences for content production.

Encyclopedic Impulse

A popular perspective on the reference work is the biography of the people who create them. The range of personality types spans a spectrum ranging from noble self-improvers to the criminally insane, though they all shared a commitment to their craft. This same commitment can be seen in present-day Wikipedians, and is a target of scorn by some critics.

As we saw with Otlet and Wells, idealists and enthusiasts are not at all uncommon in the roster of those concerned with collecting knowledge. Suzanne Briet, a contemporary of Otlet and Wells, highlighted the importance of "altruism" and "zeal in research" among the "signs of the extroverted attitude of the documentalist."[91] The famous eighteenth-century romanticist Samuel Taylor Coleridge concocted a (failed) scheme with friends for Pantisocracy, a commune in the Americas, and *Metropolitana*, an encyclopedia organized according to the branches of human knowledge rather than alphabetically.[92] (Project Xanadu was named in honor of Coleridge's poem "Kubla Khan.") And Frederick James Furnivall, a founding personality behind the nineteenth-century *Oxford English Dictionary* (OED) was known as an agnostic, vegetarian, and Socialist—characteristics for which many thought him foolish.[93]

But perhaps the most well-known personality is also one of the most tragic. Simon Winchester's *The Professor and the Madman* is the story of the *OED* and one of its most fecund contributors, Dr. William Minor. It is not clear what caused Minor's paranoid delusions, which eventually drove him to murder an innocent he mistook for the phantasms that tormented him in the night. Yet Winchester argues that Minor's devotion to the project—Minor submitted 10,000 citation slips to the *OED* documenting the early usage of terms—was perhaps one of his few solaces: partially replacing his paranoid compulsions with a constructive one that gave Minor some sense of purpose and connection to others.[94]

Regardless of whether these men were self-improvers or madmen, their passion and commitment is aptly characterized by Thomas McArthur in his history of reference works:

> In this they epitomize an important element in the history and psychology of reference materials: the passionate individuals with the peculiar taste for the hard labor of sifting, citing, listing and defining. In such people the taxonomic urge verges on the excessive. Thus, the wife of the Elizabethan lexicographer Thomas Cooper grew to fear that too much compiling would kill her husband. To prevent this, she took and burned the entire manuscript upon which he was working. Somehow, Cooper absorbed the loss—and simply sat down and started all over again.[95]

An early example of such diligence is that of Pliny the Elder's thirty-seven volume *Natural History*, one of the earliest European encyclopedias. A respected Roman admiral, statesman, and author, Pliny wrote his work of 20,000 facts with a genteel diligence. His nephew and protégé, Gaius

Plinius Cecilius Secundus, better known as Pliny the Younger, wrote to a friend of his uncle's habit of devoting every spare moment to his studies in which he would take notes while a servant read:

> I remember once his reader having mis-pronounced a word, one of my uncle's friends at the table made him go back to where the word was and repeat it again; upon which my uncle said to his friend, "Surely you understood it?" Upon his acknowledging that he did, "Why then," said he, "did you make him go back again? We have lost more than ten lines by this interruption." Such an economist he was of time![96]

Pliny even recorded his uncle's chastisement for wasting hours in walking about Rome instead of being carried in a litter from within which he could continue his studies.

Wikipedians can be a similarly compulsive and eccentric lot. So much so that some refer to themselves as Wikipediholics with a case of editcountitis, "a serious disease consisting of an unhealthy obsession with the number of edits you have made to Wikipedia."[97] One's edit count is a sort of coin of the realm. Although it is acknowledged as an arbitrary number (i.e., some might save a Wikipedia page after every tweak, whereas others may edit "offline" and paste it back when done in a single edit), one's count is a rough approximation of one's involvement and commitment to the project. In the 2006 Wikimedia Board elections only those with 400 edits could participate; in 2008 the requirement was raised to 600 edits.[98] The "Deceased Wikipedians" article states: "Please do not add people to this list who were never an integral part of the community. People in this list should have made at least several hundred edits or be known for substantial contributions to certain articles."[99]

But wait, a list of deceased Wikipedians? Indeed. Historically many reference-work contributors driven by the encyclopedic impulse also recognized that their passion would not bring them great rewards or fame. As Samuel Johnson wrote in his preface to *A Dictionary of the English Language*, "Every other author may aspire to praise; the lexicographer can only hope to escape reproach, and even this negative recompense has been yet granted to very few."[100] So, in this small way, deceased Wikipedians are acknowledged. And the list also gives a flavor of the character of Wikipedia itself. A consequence of subsuming one's self in a reference work is an appreciation of the personalities and preoccupations of those behind the seemingly staid resource. When A. J. Jacobs undertook the immense task of reading the whole *Britannica* he concluded that among the best ways to get one's own entry was to be beheaded, explore the Arctic, get castrated, design a font, or become a mistress to a monarch.[101] These were seemingly popular topics among *Britannica* editors. Similarly, the lists of Wikipedia give a similar sense of the tastes of its contributors. The "List of Lists of Lists" is one article among a dozen that were nominated as the weirdest of Wikipedia pages, giving a skewed but amusing perspective. Other weird articles included: "List Of Fictional Expletives," "Heavy Metal Umlaut," "List Of Songs Featuring Cowbells," "List Of Strange Units Of Measurement," "Professional Farter," "List Of Problems Solved By MacGyver," "Spork," "Navel Lint," "Exploding Whale," and "Twinkies in Popular Culture."[102]

Whereas tens of thousands of Wikipedians make a handful of changes, many do much more than this: a 2007 survey reported respondents averaged 8.27 hours per week on the site.[103] Some go even beyond this. For example, in 2006 the Canadian *Globe and Mail* profiled Simon Pulsifer, a Canadian in his mid-twenties at the time, who had edited more than 78,000 articles, two to three thousand of which he created.[104] How does such a habit form? Andrew Lih, author and fellow Wikipedia researcher, referred me to the story of "the red dot guy," Seth Ilys, who tells of his slip into the work as follows:

> Sometime early in 2004, I made a dot-map (example) showing the location of my hometown: Apex, North Carolina. Then I decided, what the heck, since I've done that and have the graphics program open, why don't I make maps for every town in the county. That afternoon, I did about a third of the state and it didn't make any sense to stop there, so, like Forrest Gump, I just kept on running. Eerily enough, other people started running, too, and before long nearly all of the User:Rambot U.S. census location articles will have maps.[105]

This indicates to me that it is not only the personality types of reference work compilers that are relevant, but also the character of the work itself. There is something about perusing, summarizing, compiling, and indexing. (I prefer to call this an "encyclopedic impulse" instead of McArthur's "taxonomic urge" to indicate a greater scope beyond classification, but I think we each mean the same thing.) Perhaps it is the focused, piecemeal but cumulative work that grabs some people and makes an "addict" of them. Or, as seen with Paul Otlet and H. G. Wells, the idea of liberating facts from the binding of a book is an enchanting one. And while the eccentricities are humorous and charming for the most part, there is a hint of distress in those who complain of staying up too late, falling behind with work, and suffering sore wrists. In 2006, the Wikipedia policy on blocking users stated, "Self-blocking to enforce a Wikiholiday or departure are specifically prohibited."[106] Again, it is somewhat funny that an administrator would block herself so as to stop editing, but it is also potentially sad. In the world of print, such a compulsion has led to theft, hoarding, and even murder as documented in Nicholas Basbanes's history of the "gentle madness" of book collectors.[107]

Critics have taken note of this personality trait too. But whereas I am more likely to view it with amusement, critics tend to be derisive, particularly when the excessive character of the individual joins with the like-minded to become a "MeetUp" or movement. Or, in a less flattering light, Charles Arthur, technology editor at the *Guardian*, observes Wikipedia, like many online activities, "show[s] all the outward characteristics of a cult."[108] This allegation of religious-like fervor is also seen in Gorman's reference to Wikipedia supporters as "the faithful."[109] Helprin characterizes the online Visigoths as an army whose vast bulk "may be just a bunch of whacked-out muppets" led by "little professors in glasses" (i.e., free culture proponent Lawrence Lessig).[110] And while Lanier prefers a more secular metaphor, he is nonetheless disdainful by referring to Wikipedians as a Maoist collective and Wikipedia as an "online fetish site for foolish collectivism."[111] Andrew Orlowski, a journalist at the *Register* and one of the earliest critics of Wikipedia, has published a series of articles documenting Wikipedia faults. Presumably referring to the response to his work, Orlowski returns to the religious theme when he notes "criticism from outside the Wikipedia camp has been rebuffed with a ferocious blend of irrationality and vigor that's almost unprecedented in our experience: if you thought Apple, Amiga, Mozilla or OS/2 fans were er, . . . passionate, you haven't met a wiki-fiddler. For them, it's a religious crusade."[112] And Wikipedia can get such criticism from both sides. Proponents of "aetherometry," a fringe (or pseudo) science, have also characterized Wikipedia as "a techno-cult of ignorance." However, in this case, Wikipedia is not being criticized for being overly populist, but for failing to recognize a "dissident science" in favor of the "power-servant peer-review institutions of Big Science."[113]

Here, the passions and eccentricities common to compilers throughout the centuries become a feature of the debate between supporters and critics themselves.

Technological Inspiration

Index cards, microfilm, and loose-leaf binders inspired early documentalists to envision greater information access. Furthermore, these technologies had the potential to change how information was thought of and handled. Otlet's monographic principle, discussed in chapter 2, recognized that with technology one would be able to "detach what the book amalgamates, to reduce all that is complex to its elements and to devote a page [or index card] to each."[114] (The incrementalism frequently alluded to in Wikipedia production is perhaps an instance of this principle in operation.) Similarly, Otlet's Universal Decimal Classification system would allow one to find these fragments of information easily. These notions of decomposing and rearranging information are again found in current Web 2.0 buzzwords such as *tagging*, *feeds*, and *mash-ups*, or the popular Apple slogan "rip, mix, and burn."[115] And critics object.

As noted, Michael Gorman did not launch his career as a Web 2.0 curmudgeon with a blog entry about Wikipedia; he began with an opinion piece in the *Los Angeles Times*. In his first attack, prompted by the "boogie-woogie Google boys" claim that the perfect search would be like "the mind of God," Gorman lashes out at Google and its book-scanning project. His concern was not so much about the possible copyright infringement of scanning and indexing books, which was the dominant focus of discussion at the time, but the type of access it provided. Gorman objects to full-text search results that permit one to peruse a few pages on the screen:

> The books in great libraries are much more than the sum of their parts. They are designed to be read sequentially and cumulatively, so that the reader gains knowledge in the reading. . . . The nub of the matter lies in the distinction between information (data, facts, images, quotes and brief texts that can be used out of context) and recorded knowledge (the cumulative exposition found in scholarly and literary texts and in popular nonfiction). When it comes to information, a snippet from Page 142 might be useful. When it comes to recorded knowledge, a snippet from Page 142 must be understood in the light of pages 1 through 141 or the text was not worth writing and publishing in the first place.[116]

From this initial missive, Gorman's course of finding fault with anything that smelled of digital populism was set, and would eventually bring him to Wikipedia. (Ironically, he became an exemplar of the successful opinion blogger: shooting from the hip, irreverent, and controversial.)

Yet others enthusiastically counter Gorman's disdain for the digital. Kevin Kelly, previously encountered in the hive-mind debate, resurrected the spirit of the monographic principle in a May 2006 *New York Times Magazine* essay about the "liquid version" of books. Instead of index cards and microfilm, the liquid library is enabled by the link and the tag, maybe "two of the most important inventions of the last 50 years."[117] Kelly noted that the ancient Library of Alexandria was evidence that the dream of having "all books, all documents, all conceptual works, in all languages" available in one place is an old one; now it might finally be realized. Despite being unaware that the curtain was raised almost a century ago, his reprise is true to Otlet's vision:

> The real magic will come in the second act, as each word in each book is cross-linked, clustered, cited, extracted, indexed, analyzed, annotated, remixed, reassembled and woven deeper into the culture than ever before. In the new world of books, every bit informs another; every page reads all the other pages. . . . At the same time, once digitized, books can be unraveled into single pages or be reduced further, into snippets of a page. These snippets will be remixed into reordered books and virtual bookshelves.[118]

It's not hard to see Wikipedia as a "reordered book" of reconstituted knowledge. Gorman, probably familiar with some of the antecedents of the liquid library, given his reference to "Universal Bibliographic Control" above and skepticism of microfilm below, considers such enthusiasm to be ill founded: "This latest version of Google hype will no doubt join taking personal commuter helicopters to work and carrying the Library of Congress in a briefcase on microfilm as 'back to the future' failures, for the simple reason that they were solutions in search of a problem."[119] Conversely, Andrew Keen fears it is a problem in the guise of a solution, claiming the liquid library "is the digital equivalent of tearing out the pages of all the books in the world, shredding them line by line, and pasting them back together in infinite combinations. In his [Kelly's] view, this results in 'a web of names and a community of ideas.' In mine, it foretells the death of culture."[120]

Yet, Kevin Drum, a blogger and columnist, notes that this dictum of sequentially reading the inviolate continuity of pages isn't even the case in the "brick-and-mortar library" today: "I browse. I peek into books. I take notes from chapters here and there. A digitized library allows me to do the same thing, but with vastly greater scope and vastly greater focus."[121] As far back as 1903 Paul Otlet felt the slavish dictates of a book's structure were a thing of the past: "Once one read; today one refers to, checks through, skims. Vita brevis, ars longa! There is too much to read; the times are wrong;

the trend is no longer slavishly to follow the author through the maze of a personal plan which he has outlined for himself and which in vain he attempts to impose on those who read him."[122] In fact, scholars have always had varied approaches to reading.[123] Francis Bacon (1561–1626) noted, "Some books are to be tasted, others to be swallowed, and some few to be chewed and digested."[124] A twelfth-century manuscript on "study and teaching" recommended that a prudent scholar "hears every one freely, reads everything, and rejects no book, no person, no doctrine," but "If you cannot read everything, read that which is more useful."[125] And as (un)usual as it may be for anyone to always read a book from start to finish, Gorman's skepticism also includes an accusation inevitable to discussions about contemporary technology: hype, or "a wonderfully modern manifestation of the triumph of hope and boosterism over reality."[126] Wikipedia critics claim that technology has inspired hyperbole. In response to the Seigenthaler's complaint about fabrications in his biographical article, Orlowski writes the resulting controversy "would have been far more muted if the Wikipedia project didn't make such grand claims for itself."[127] Nick Carr writes that what "gets my goat about Sanger, Wales, and all the other pixel-eyed apologists for the collective mediocritization of culture" is that they are "all in the business of proclaiming the dawn of a new, more perfect age of human cognition and understanding, made possible by the pulsing optical fibers of the internet."[128] Jaron Lanier, coiner of the term *Digital Maoism*, concurs: "the problem is in the way the Wikipedia has come to be regarded and used; how it's been elevated to such importance so quickly."[129] Building on Lanier, Gorman speaks to the hype, and many of his other criticisms:

> Digital Maoism is an unholy brew made up of the digital utopianism that hailed the Internet as the second coming of Haight-Ashbury—everyone's tripping and it's all free; pop sociology derived from misreading books such as James Surowiecki's 2004 *The Wisdom of Crowds: Why the Many are Smarter Than the Few and How Collective Wisdom Shapes Business, Economies, Societies, and Nations*; a desire to avoid individual responsibility; anti-intellectualism—the common disdain for pointy headed professors; and the corporatist "team" mentality that infests much modern management theory.[130]

Helprin likens Wikipedia to the *Great Soviet Encyclopedia* wherein the Kremlin sent out doctored photographs and updated pages to rewrite history: "Revision as used by the Soviets was a tool to disorient and disempower the plasticized masses. Revision in the wikis is an inescapable attribute that eliminates the fixedness of fact. Both the Soviets and the wiki builders imagined and imagine themselves as attempting to reach the truth."[131] Likewise, Carr continues his criticism by noting: "Whatever happens between Wikipedia and Citizendium, here's what Wales and Sanger cannot be forgiven for: They have taken the encyclopedia out of the high school library, where it belongs, and turned it into some kind of totem of 'human knowledge.' Who the hell goes to an encyclopedia looking for 'truth,' anyway?"[132]

Of course, one must ask to what extent has Wikipedia made "such grand claims for itself"? As I belabored in my discussions about NPOV, Wikipedia has few, if any, pretensions to "truth." As is stressed in the "Verifiability" policy, "The threshold for inclusion in Wikipedia is verifiability, not truth—that is, whether readers are able to check that material added to Wikipedia has already been published by a reliable source, not whether we think it is true."[133] Unlike the launching of the *Third*, there was no ill-conceived press release claiming Wikipedia to be truth incarnate. Furthermore, encyclopedias gained their present shine of truth when they were first sold to schools in the middle of the twentieth century. Also, we must remember Wikipedia was not started with the intention of creating a Maoistic hive intelligence. Rather, Nupedia's goal was to produce an encyclopedia that could be available to—not produced by—anyone. When the experiment of allowing anyone to edit on a complementary wiki succeeded beyond its founders' expectations, and Wikipedia was born, two things happened. First, journalists, and, later, popular-press authors, seized upon its success as part of a larger theory about technology-related change. For example, Don Tapscott and Anthony Williams reference the wiki phenomenon in the title of their book *Wikinomics;*[134] they use a brief account of Wikipedia to launch a much larger case of how businesses should learn from and adapt their strategies to new media and peer collaboration. In *Infotopia* Cass Sunstein engages the Wikipedia phenomenon more directly, and identifies some strengths of this type of group decision making and knowledge production, but also illuminates possible faults.[135] Using Wikipedia as a metaphor has become so popular that Jeremy Wagstaff notes that comparing something to Wikipedia is "The New Cliche": "You know something has arrived when it's used to describe a phenomenon. Or what people hope will be a phenomenon."[136] Second, as seen earlier, Wikipedians themselves sought to understand how the experiment turned out so well and engaged in discussions about whether those larger theories applied.

However, at the launch of Wikipedia, Ward Cunningham, Larry Sanger, and Jimmy Wales all expressed some skepticism regarding its success as an encyclopedia, a conversation that continued among Wikipedia supporters until at least 2005.[137] And as evidence of early modesty, consider the following message from Sanger at the start of Wikipedia:

> Suppose that, as is perfectly possible, Wikipedia continues producing articles at a rate of 1,000 per month. In seven years, it would have 84,000 articles. This is entirely possible; Everything2, which uses wiki-like software, reached 1,000,000 "nodes" recently.[138]

Some thought this was a stretch. In 2002, online journalist Peter Jacso included Wikipedia in his "picks and pan" column: he "panned" Wikipedia by likening it to a prank, joke, or an "outlet for those who pine to be a member in some community." Jacso dismissed Wikipedia's goal of producing 100,000 articles with the comment: "That's ambition," as this "tall order" was twice the number of articles in the sixth edition of the *Columbia Encyclopedia.*[139] Yet, in September 2007, shy of its seven-year anniversary, the English Wikipedia had two million articles (over twenty times Sanger's estimate), proving that making predictions about Wikipedia is definitely a hazard—prompting betting pools on when various million-article landmarks will be reached.[140]

Granting that technology pundits make exaggerated claims, but not always to the extent that critics allege, prominent Wikipedians tend to be more moderate in their claims: in response to the Seigenthaler incident in 2005 Wales cautioned that while they wanted to rival *Britannica* in quantity and quality, that goal had not yet been achieved and that Wikipedia was "a work in progress."[141] The Wikipedia article "What It Is Not" disclaims many of the labels commonly attributed to it, including that it is "not an anarchy."[142] And of the ten things you might "not know about Wikipedia":

> We do not expect you to trust us. It is in the nature of an ever-changing work like Wikipedia that, while some articles are of the highest quality of scholarship, others are admittedly complete rubbish. We are fully aware of this. We work hard to keep the ratio of the greatest to the worst as high as possible, of course, and to find helpful ways to tell you in what state an article currently is. Even at its best, Wikipedia is an encyclopedia, with all the limitations that entails. It is not a primary source. We ask you not to criticize Wikipedia indiscriminately for its content model but to use it with an informed understanding of what it is and what it isn't. Also, as some articles may contain errors, please do not use Wikipedia to make critical decisions.[143]

While pundits might seize upon Wikipedia as an example of their argument of dramatic change, most Wikipedia supporters tend to express more surprise than hyped-up assuredness. In response to the Seigenthaler incident in 2005, the British newspaper the *Guardian* characterized Wikipedia as "one of the wonders of the internet":

In theory it was a recipe for disaster, but for most of the time it worked remarkably well, reflecting the essential goodness of human nature in a supposedly cynical world and fulfilling a latent desire for people all over the world to cooperate with each other without payment. The wikipedia is now a standard source of reference for millions of people including school children doing their homework and postgraduates doing research. Inevitably, in an experiment on this scale lots of entries have turned out to be wrong, mostly without mal-intent. . . . Those who think its entries should be taken with a pinch of salt should never forget that there is still plenty of gold dust there.[144]

Economist and author John Quiggin notes: "Still, as Bismarck is supposed to have said 'If you like laws and sausages, you should never watch either one being made.' The process by which Wikipedia entries are produced is, in many cases, far from edifying: the marvel, as with democracies and markets, is that the outcomes are as good as they are."[145] Bill Thompson, BBC digital culture critic, wrote "Wikipedia is flawed in the way Ely Cathedral is flawed, imperfect in the way a person you love is imperfect, and filled with conflict and disagreement in the way a good conference or an effective parliament is filled with argument."[146] The same sentiment carried through in many of the responses to Jaron Lanier's "Digital Maoism" article. Yochai Benkler replies, "Wikipedia captures the imagination not because it is so perfect, but because it is reasonably good in many cases: a proposition that would have been thought preposterous a mere half-decade ago."[147] Science fiction author and prominent blogger Cory Doctorow writes: "Wikipedia isn't great because it's like the *Britannica*. The *Britannica* is great at being authoritative, edited, expensive, and monolithic. Wikipedia is great at being free, brawling, universal, and instantaneous."[148] Kevin Kelly, proponent of the hive mind and liquid library, responds that Wikipedia surprises us because it takes "us much further than seems possible. . . . because it is something that is impossible in theory, and only possible in practice."[149]

And Wikipedia defenders are not willing to cede the quality ground altogether. On December 14, 2005, the prestigious science journal *Nature* reported the findings of a commissioned study in which subject experts reviewed forty-two articles in Wikipedia and *Britannica*; it concluded "the average science entry in Wikipedia contained around four inaccuracies; Britannica, about three."[150] Of course, this catered to the interests of *Nature* readers and a topical strength of Wikipedia contributors. Wikipedia may not have fared so well using a random sampling of articles or on humanities subjects. Three months later, in March 2006, *Britannica* boldly objected to the methodology and conclusions of the *Nature* study in a press release and large ads in the *New York Times* and the *London Times*. Interestingly, by this time, Wikipedia had already fixed all errors identified in the study—in fact they were corrected within three days of learning of the specific errors.[151]

Yet the critics don't accept even this more moderated appreciation of Wikipedia as being imperfect but surprisingly good. Orlowski writes such sentiments are akin to saying: "Yes it's garbage, but it's delivered so much faster!"[152] In a widely read article on Wikipedia for the *The New Yorker*, Stacy Schiff reported Robert McHenry, former editor in chief of the *Encyclopædia Britannica*, as saying "We can get the wrong answer to a question quicker than our fathers and mothers could find a pencil."[153] Carr is willing to concede a little more, but on balance still finds Wikipedia lacking:

> In theory, Wikipedia is a beautiful thing—it has to be a beautiful thing if the Web is leading us to a higher consciousness. In reality, though, Wikipedia isn't very good at all. Certainly, it's useful—I regularly consult it to get a quick gloss on a subject. But at a factual level it's unreliable, and the writing is often appalling. I wouldn't depend on it as a source, and I certainly wouldn't recommend it to a student writing a research paper.[154]

Furthermore, whereas Wikipedia supporters see "imperfect" as an opportunity to continue moving forward, critics view user-generated content as positively harmful: that "Misinformation has a negative value," or that "what is free is actually costing us a fortune."[155] (Perhaps this is a classical case of perceiving a glass to be either half empty or half full.) Or, much like the enormously popular parody of an inspirational poster that declared "Every time you masturbate, God kills a kitten," Keen concludes: "Every visit to Wikipedia's free information hive means one less customer for professionally researched and edited encyclopedia such as Britannica."[156] And Carr fears that using the Internet to pursue (suspect) knowledge is actually "making us stupid."[157]

Although technology can inspire, it can cause others to despair. For some, like Gorman's dismissal of the Library of Congress in a briefcase, the technology may inspire nothing but a "back to the future" failure. For others, like Keen, the proclaimed implications of the technology are real. Yet, whereas Anderson loves Rhapsody, the online music service, Keen has lost Tower Records, the defunct brick-and-mortar store. Here we can observe a generality of history: change serves some better than others. On this point these arguments seem like those of any generational gap, as Gorman points out:

> Perceived generational differences are another obfuscating factor in this discussion. The argument is that scholarship based on individual expertise resulting in authoritative statements is somehow passé and that today's younger people think and act differently and prefer collective to individual sources because of their immersion in a digital culture. This is both a trivial argument (as if scholarship and truth were matters of preference akin to liking the Beatles better than Nelly) and one that is demeaning to younger people (as if their minds were hopelessly blurred by their interaction with digital resources and entertainments).[158]

Nonetheless, Gorman manages to sound like an old man shaking his fist when he complains that "The fact is that today's young, as do the young in every age, need to learn from those who are older and wiser."[159] Clay Shirky summarizes Gorman's position from the perspective of the new generation: "according to Gorman, the shift to digital and network reproduction of information will fail unless it recapitulates the institutions and habits that have grown up around print."[160] Scott McLemee, a columnist at *Inside Higher Ed,* more amusingly notes: "The tone of Gorman's remedial lecture implies that educators now devote the better part of their day to teaching students to shove pencils up their nose while Googling for pornography. I do not believe this to be the case. (It would be bad, of course, if it were.)"[161]

Finally, some of this conflict might be characterized as "much ado about nothing." Both *Webster's Third* and Wikipedia have attracted a fair amount of punditry: reference works are claimed as proxies and hostages in larger battles, and I suspect some of the combatants argue for little other than their own self-aggrandizement. When reading generational polemics I remind myself of Douglas Adams's humorous observation that everything that existed when you were born is considered normal and you should try to make a career out of anything before your thirtieth birthday as it is thought to be "incredibly exciting and creative." Of course, anything after that is "against the natural order of things and the beginning of the end of civilisation as we know it until it's been around for about ten years when it gradually turns out to be alright really." Even so, with every generation we undergo a new round of "huffing and puffing."[162] This is because "old stuff gets broken faster than the new stuff is put in its place," as Clay Shirky notes in a blog entry about the collapse of print journalism. Or, as hypothesized by Steve Weber in his study of open source, the stridency of critics arises because it is easier to see "what is going away than what is struggling to be born" but that there can be a positive side to "creative destruction" if we are sufficiently patient.[163]

Conclusion

Reference works can prompt and embody currents of social unease. As seen in Morton's history of *Webster's Third*, much of the controversy associated with its publication was about something other than the merits of that particular dictionary. I generalize the argument by briefly looking to the past for how reference works have been involved in a larger conservative versus progressive tension, and by asking how Wikipedia might be entangled in a similar debate today.

On this point, the conversation about Wikipedia can be understood with respect to a handful of themes. Clearly, the way in which content is produced has changed. It is not surprising that people question whether this type of collaboration is good, bad, or could be improved upon in any case. Furthermore, Wikipedia is an (imperfect) realization of a long-pursued vision for a universal encyclopedia. This vision is challenged by critics as an unlikely utopia, or a dangerous dystopia. Also, how to make sense of the sometimes rancorous character of the discussion? We might understand the doggedness of some of the supporters and critics in light of an encyclopedic impulse and the longer history of bibliophilic passion. Central to the discussion is also a long-debated question about technology and change: although technology may inspire some toward a particular end, it might also disgust others and effect changes that are not welcome. Ultimately, I find a reasoned middle ground to be most compelling. In a keynote speech before the Association of Research Libraries, Hunter R. Rawlings III, classics scholar and former president of Cornell, noted that we should not confuse the useful measures of relevance and popularity in finding information with the means to validate it; we must continue to develop means of "critical judgment."[164] Wikipedia can serve not only as a reference work, but also, at the same time, as a study of how knowledge is constructed and contested.

8 Conclusion: "A Globe in Accord"

The problem with Wikipedia is that it only works in practice. In theory, it can never work.
—Zeroeth Law

Everyone who comes across Raul's laws eventually adds one of their own.
—Ben's Revolting Realization

At Wikimania 2007, a gathering of Wikimedia contributors in Taipei, one of the free gifts received during registration was a spherical puzzle. Like any other jigsaw, the pieces must be fit together, but in this case they form a globe much like the one seen near the top of every Wikipedia article. The Wikipedia logo is that of an incomplete world of characters, each piece representing a different language. In discussing Wikipedia's culture, I use the metaphor of a puzzle to explain the ways in which neutral point of view and good faith complement each other in the collaborative production of an encyclopedia. NPOV makes it possible for the jigsaw shapes to actually be fitted together, and good faith facilitates the process—sometimes frustrating, sometimes fun—of putting them together with one's peers. In accordance with Ben's Revolting Realization at the head of this chapter, this idea—and this book itself I suppose—is my own addendum to the Laws of Wikipedia.

But this metaphor of a jigsaw puzzle is even more appropriate when I think back to H. G. Wells and his "world brain." (This occurred to me at five a.m. on the last day of Wikimania as I gazed unfocused at the puzzle box sitting on the nightstand next to the bed.) Wells and others pursuing the vision of a universal encyclopedia had hoped that new technologies, be they index cards and microfilm or computer networks, might somehow address the difficult puzzle of the world's troubles. Even if more recent visionaries aren't quite as utopian—or perhaps naïve—as Wells and Otlet were, there is a hopeful and global aspiration nonetheless.[1] In fact, the motto of Wikimania 2007 was "a Globe in Accord"—and I was struck by the sight of multilingual participants wearing "I speak" badges enumerating the languages in which they could converse and help.

However, just as "neutral" should not be understood as a description of the encyclopedia but as an aspiration and intentional stance of its contributors, one should appreciate ideals of universalism, openness, and good faith in a similar light. For example, there are inherent tensions (e.g., "the tyranny of structurelessness") and practical difficulties (e.g., Wikipedia office actions) within an open content community. Similarly, if one were to read my focus on good faith (assuming the best of others, striving for patience, civility, and humor) as implying that Wikipedia is a harmonious community of benevolent saints, one would be wrong.

If forced to simplify the complexities of online community by way of a single theory I would resort to Godwin's Law, first observed on Usenet.[2] We often see the world in the parochial terms of "us versus them," and we tend to be less favorable in judging others than we judge ourselves—and then we are amazingly adept at justifying and rationalizing our own mistakes.[3] Given the lack of social context in online interactions (distant, nearly anonymous, and transitory), it is not surprising that people sometimes end up calling each other Nazis. This is why when Wikipedia began to experience its first serious growing pains Wales called for a "culture of co-operation" unlike the "culture of conflict embodied in Usenet."[4] And although Wikipedia might be "dedicated to a higher good," I agree with journalist Stacy Schiff that "it is also no more immune to human nature than any other utopian project. Pettiness, idiocy, and vulgarity are regular features of the site. Nothing about high-minded collaboration guarantees accuracy, and open editing invites abuse."[5] What Wikipedia's collaborative culture does, what any culture with positive norms like "Don't Bite the Newcomers" or "Assume Good Faith" can do, is dampen Godwin's Law and call upon "the better angels of our nature."[6] Those pursuing the universal encyclopedia believe that while our better nature is not always present, it is at least latent. For example, in response to social arguments about "survival of the fittest" arising from Darwin's *The Origin of Species*, Peter Kropotkin, anarchist and contributor to the 1911 *Encyclopædia Britannica*, wrote "Mutual aid is as much a law of animal life as mutual struggle."[7] There are even times when we can surprise ourselves, such as when thousands of (previous) strangers come together to build a world encyclopedia. As Sanger notes, to build a universal encyclopedia one doesn't need "faith in the possibility of knowledge" but in "human beings being able to work together."[8] The question, then, is how is such a thing possible? Or as Peter Kollock wrote about cooperative online efforts before Wikipedia: "For a student of social order, what needs to be explained is not the amount of conflict but the great amount of sharing and cooperation that does occur in online communities."[9]

One's first impulse in answering the question about Wikipedia's success is to focus on technology. Clearly, as is apparent in my history, technology has played a significant role in inspiring the vision of a universal encyclopedia. And beyond inspiration, networking technology and its related collaborative techniques can enable openness and accessibility, furthering accountability and the socialization of newcomers. On a wiki, contributors can communicate asynchronously and contribute incrementally. Tasks can be modularized. Changes are easily reverted. Accessible documentation, discussion pages, templates, and automated tools further coordination. However, technology, while important, is insufficient. Plenty of projects fail despite the wiki pixie dust. This is why the question "How is something like Wikipedia possible?" leads me to the question "How can we understand Wikipedia's collaborative culture?" As noted, Larry Sanger concedes that at the start he mistakenly "denied that Wikipedia was a community, claiming that it was, instead, only an encyclopedia project."[10] This is a type of mistake he thinks others now make with respect to technology:

> It is not anything magic about wiki software in particular that makes Wikipedia work as well as it does. Wikipedia's success is more due to the fact that it is strongly collaborative than that it is a wiki. Wikis and the Wikipedia model are one way to enable strong collaboration, but they are not only one way. I think that the Wikipedia community made a mistake when it decided that it's the wiki part that explained Wikipedia's success.[11]

Perhaps a lot of the criticism against Wikipedia and "Web 2.0" relates to this issue. People seize upon *wiki* as a buzzword, implying they can magically transform business, government, or anything really. Observing this hyped rendering of technology, some critics ask, but what of individual difference and social bonds? Wikipedia supporters argue these things have been there all along. This is why a focus on community and culture are necessary to understanding Wikipedia; as Sanger notes, "while collaborative systems should be designed with the needs and values of participants in mind, I think that a certain culture or set of values, is necessary in order to make collaboration work."[12] My argument is that good faith social norms (combined with wiki features) constructively facilitate Wikipedia collaboration. However, more autocratic forms of authority may be necessary to defend against those acting in bad faith or when there is no community consensus. Hence, egalitarian open content communities are sometimes (ironically) led by a "benevolent dictator." Jimmy Wales serves in this role at Wikipedia and has influenced much of its culture. Yet, if such leadership or institutional governance persistently fails, the community might then fork.

Even if one accepts my argument about the importance of culture, some might argue my portrayal is off the mark. I've already qualified my focus on good faith as an aspiration and cultural norm rather than a description of all Wikipedia practice. (Though the corpus of norms and their imperfect implementation is remarkable still.) Yet some readers might claim things have changed at Wikipedia: it may have once been an encyclopedia with potential, been produced by an open content community, or had a culture of good faith, but not now.

Wikipedia's status as an encyclopedia was debated from the start, even by its founders, and continues to be thought suspect by critics, particularly when a new scandal erupts as they seem to do every so often. This then prompts much discussion. In fact, the community has discussed every conceivable aspect of its identity and work. As I noted at the beginning of this book, this conversation is frequently exasperating and often humorous, but we now know it is also rather pragmatic and governed by good faith norms. Indeed, Wikipedia is an exemplar of the reflective character of open content communities. And just when arguments that Wikipedia would never amount to anything ceased, new arguments about its death took their place. Based on research showing that Wikipedia contribution is slowing, journalist Stephen Foley asks, "is Wikipedia cracking up?"[13] In 2005, law professor Eric Goldman predicted Wikipedia would "fail" in 2010 (i.e., close access or become spam ridden), repeated the prediction in 2006, and in 2009 made the claim at a conference.[14] (If you can still edit Wikipedia when you read this book, it is safe to conclude that he was wrong.)

No doubt, the community will change, but change is inevitable—and my efforts are necessarily fixed in a particular slice of time. Also, "golden years" tend to be subjective and relative. I began this work in 2004, the same year a self-described "old-timer" mentioned he began his wiki career and the same year in which another (older old-timer) told me the project began to go downhill. (JDG's First Law notes each "wave or generation" of Wikipedia editors corresponds to "the human gestation period," which means about nine months.[15]) I too have concerns about Wikipedia's quality, community, and culture as it evolves. And just like any community Wikipedia does change. It has been relatively successful and has faced extraordinary growing pains. Almost a century ago the seminal sociologist Max Weber noted that organizations often develop toward bureaucratic forms. We shouldn't be surprised that the same has happened to Wikipedia; perhaps those who are disenchanted should think of themselves as "wiki entrepreneurs," preferring the fast and flexible environment of a small community. And, as Weber notes, "When those subject to bureaucratic control seek to escape the influence of the existing bureaucratic apparatus, this is normally possible only by creating an organization of their own which is equally subject to bureaucratization."[16] It seems as if Weber was speaking of forking over a century ago.

In fact, I considered those who have left Wikipedia to begin anew at another wiki as part of its legacy. In the most extreme and unlikely case, even if the community disappeared and all that was left was a snapshot of its content, Wikipedia still would have been an amazing phenomenon. Among all of those individuals throughout history who have pursued the vision of a universal encyclopedia, Wikipedians have come closest to realizing it. Even the lifeless remains of Wikipedia content would continue to be a useful resource. And there would be dozens of projects with former Wikipedians still pursuing the vision of accessible knowledge and the joy of collaborating in good faith. However, Wikipedia has always been a puzzle. Born almost as a happy accident, growing far beyond anyone's expectations, and applauded not because it is perfect but because it is confoundingly good, I expect Wikipedia will continue to surprise us.

Notes

Bibliographic Note

Most of the sources used in this book are Web pages easily accessible to the reader. Because I cite many Wikimedia related pages, I elide some information from their preferred format for brevity: I use "Wikipedia" as the author and publisher, rather than "Wikipedia contributors" and "Wikipedia, The Free Encyclopedia" respectively. Given the large amount of grammatical and syntactical deviations online, and that some sources are not native English speakers, I mostly present quoted text verbatim, with minimal corrections or editorial caveats such as "[sic]." In the printed book I do not provide URLs for printed or easily locatable sources. For online sources, where available, the versioned URL (i.e., a "stable" or "permanent" link) is used. I sometimes reference different versions of the same wiki page or older versions so as to provide a historical source, for comparison, or because I noted text that might have been subsequently edited—wikis do change—but I expect that its spirit remains the same.

However, I recommend consulting the online bibliography where I include hypertextual references whenever possible, including for printed sources now online. This is common for recent works, sometimes as preprints or author's copies, and for older works in the public domain; I use the publication date of the version I used. For older works, if necessary, I often mention the original publication date. Page numbers might refer to the original pagination or to an online printout; I hope it will be clear to the reader which is the case.

Chapter 1: Nazis and Norms

1. AndyL, "Neo-Nazis Attempt to Stack Wikivote on Jewish," wikien-l, February 6, 2005, http://marc.info/?i=BE2BEA7C.4E78%andyl2004@sympatico.ca (accessed February 6, 2005).

2. AndyL, "A Neo-Nazi Wikipedia," wikien-l, August 19, 2005, http://marc.info/?i=20050819221529.ZQFS2134.tomts48-srv.bellnexxia.net@[209.226.175.82] (accessed August 23, 2005).

3. Mike Godwin, "Meme, Counter-Meme," *Wired* 2, no. 10 (October 1994).

4. Wikipedia, "User:Raul654/Raul's Laws," Wikipedia, July 10, 2009, http://en.wikipedia.org/?oldid=301373968 (accessed July 20, 2009).

5. Jimmy Wales, "Re: A Neo-Nazi Wikipedia," wikien-l, August 23, 2005, http://marc.info/?i=430B6975.9040906@wikia.com (accessed August 23, 2005).

6. Wikipedia, "Wikipedia:Etiquette," Wikipedia, October 23, 2008, http://en.wikipedia.org/?oldid=247072399 (accessed November 3, 2008); Wikipedia, "Wikipedia:Assume Good Faith," Wikipedia, November 1, 2008, http://en.wikipedia.org/?oldid=248943513 (accessed November 3, 2008); Wikipedia, "Wikipedia:Please Do Not Bite the Newcomers," Wikipedia, October 16, 2008, http://en.wikipedia.org/?oldid=245559838 (accessed November 3, 2008).

7. Wikimedia Foundation, "Vision," Wikimedia, September 1, 2007, http://wikimediafoundation.org/wiki/Vision (accessed June 5, 2008).

8. H. G. Wells, "The Idea of a World Encyclopedia," *Nature* 138 (November 28, 1936): 920–921.

9. Liane Gouthro, "Building the World's Biggest Encyclopedia," *PC World* (March 10, 2000).

10. Jimmy Wales, "Historical Draft on Digital Encyclopedias," email message to author, June 1, 2006.

11. Ward Cunningham, "Correspondence on the Etymology of Wiki," November 2003, http://c2.com/doc/etymology.html (accessed October 4, 2008).

12. Wikipedia, "Help:Category," Wikipedia, December 1, 2008, http://en.wikipedia.org/?oldid=255169078 (accessed December 6, 2008); Wikipedia, "Category:1122 Births," Wikipedia, October 22, 2008, http://en.wikipedia.org/?oldid=247041224 (accessed December 6, 2008).

13. Wikipedia, "Help:Template," Wikipedia, October 25, 2008, http://en.wikipedia.org/?oldid=247625503 (accessed December 5, 2008); Wikimedia, "Help:A Quick Guide to Templates," Wikimedia, June 29, 2008, http://meta.wikimedia.org/?oldid=1063186 (accessed December 5, 2008).

14. Wikipedia, "Template:Pp-Vandalism," Wikipedia, September 5, 2008, http://en.wikipedia.org/?oldid=236387424 (accessed December 6, 2008); Wikipedia, "Category:Protected Against Vandalism," Wikipedia, November 2, 2008, http://en.wikipedia.org/?oldid=249177947 (accessed December 6, 2008).

15. Despite its original simplicity, the technical and cultural complexity of Wikipedia has increased with its popularity; newcomers can now acclimate themselves with two popular-press books: John Broughton, *Wikipedia: The Missing Manual* (Sebastopol, CA: Pogue Press, 2008); Charles Matthews, Ben Yates, and Phoebe Ayers, *How Wikipedia Works* (San Francisco: No Starch Press, 2008).

16. Wikipedia, "History of Wikipedia," Wikipedia, January 8, 2009, http://en.wiki pedia.org/?oldid=262685944 (accessed January 8, 2009).

17. Wikipedia, "Wikipedia:About," Wikipedia, August 12, 2009, http://en.wikipedia .org/?oldid=307641830 (accessed August 17, 2009).

18. Larry Sanger, "Translation," nupedia-l, March 20, 2000, http://web.archive.org /web/20030516085026/www.nupedia.com/pipermail/nupedia-l/2000 -March/000064.html (accessed June 7, 2006).

19. Alexander Halavais and Derek Lackaff, "An Analysis of Topical Coverage of Wikipedia," *Journal of Computer-Mediated Communication* 13 (2008): 429–440; Aniket Kittur, Ed H. Chi, and Bongwon Suh, "What's in Wikipedia? Mapping Topics and Conflict Using Socially Annotated Category Structure," in *CHI '09: Proceedings of the 27th International Conference on Human Factors in Computing Systems* (ACM, 2009), 1509–1512; Anselm Spoerri, "What Is Popular on Wikipedia and Why?" *First Monday* 12, no. 4 (April 2007).

20. Wikipedia, "Wikipedia:100,000 Feature-Quality Articles," Wikipedia, March 13, 2009, http://en.wikipedia.org/?oldid=276882634 (accessed May 21, 2009); Wikipe dia, "Wikipedia:Version 1.0 Editorial Team/Assessment," Wikipedia, May 14, 2009, http://en.wikipedia.org/?oldid=289898436 (accessed May 21, 2009).

21. Jim Giles, "Internet Encyclopaedias Go Head to Head," *Nature* (December 14, 2005).

22. George Bragues, "Wiki-Philosophizing in a Marketplace of Ideas: Evaluating Wikipedia's Entries on Seven Great Minds," Social Science Research Network, April 2007, http://ssrn.com/abstract=978177 (accessed March 3, 2009), 14, 46.

23. Wikipedia, "Gowanus, Brooklyn," Wikipedia, May 3, 2008, http://en.wikipedia .org/?oldid=209839170 (accessed June 3, 2008).

24. Wikipedia, "Wikipedia:Editing Frequency," Wikipedia, November 30, 2008, http://en.wikipedia.org/?oldid=254994395 (accessed December 1, 2008).

25. Aniket Kittur, Bongwon Suh, Bryan A. Pendleton, and Ed H. Chi., "He Says, She Says: Conflict and Coordination in Wikipedia," in *CHI 2007 Proceedings: Online Representation of Self* (ACM, 2007).

26. Jimmy Wales, "Re: Wikipedia, Emergence, and the Wisdom of Crowds," wikipedia-l, May 4, 2005, http://marc.info/?i=4278DF44.1030905@wikia.com (accessed May 4, 2005). For discussion of what constitutes a contribution, see Aaron Swartz, "Who Writes Wikipedia?" Aaron Swartz's Raw Thought blog, September 4, 2006, http://www.aaronsw.com/weblog/whowriteswikipedia (accessed September 4, 2006); Aniket Kittur, Ed Chi, Bryan A. Pendleton, Bongwon Suh, and Todd Mytkowicz, "Power of the Few Vs. Wisdom of the Crowd: Wikipedia and the Rise of the Bourgeoisie," in *25th Annual ACM Conference on Human Factors in Computing Systems (CHI 2007)* (New York: ACM, 2007); Reid Priedhorsky, Jilin Chen, Shyong Lam, Kathering

Panciera, Loren Terveen, and John Riedl, "Creating, Destroying, and Restoring Value in Wikipedia," in *Proceedings of GROUP 2007* (New York: ACM, 2007); Felipe Ortega and Jesus Gonzalez-Barahona, "On the Inequality of Contributions to Wikipedia," in *Proceedings of the 41st Hawaiian International Conference on System Sciences (HICSS 2008)* (2008). For changes in contribution patterns, see Philip Ball, "The More, the Wikier: the Secret to the Quality of Wikipedia Entries Is Lots of Edits by Lots of People," *Nature News* (February 27, 2007).

27. Wikipedia, "Wikipedia:List of Administrators," Wikipedia, August 11, 2009, http://en.wikipedia.org/?oldid=307325114 (accessed August 11, 2009).

28. Walter, "The English Language Edition of Wikizine," May 21, 2009, http://en.wikizine.org/ (accessed May 21, 2009).

29. Wikipedia, "Wikipedia:WikiProject," Wikipedia, May 6, 2009, http://en.wikipedia.org/?oldid=288185213 (accessed May 21, 2009); ClockworkSoul, "Igor's All Projects List," Wiki-Igor, August 3, 2009, http://wiki-igor.net/all-projects.jsp (accessed August 3, 2009).

30. Ruediger Glott, Philipp Schmidt, and Rishab Ghosh, "Wikipedia Survey—First Results," April 9, 2009, http://upload.wikimedia.org/wikipedia/foundation/a/a7/Wikipedia_General_Survey-Overview_0.3.9.pdf (accessed April 24, 2009).

31. For *Wikizine, Wikipedia Signpost,* and "Village Pump," see Walter, "The English Language Edition of Wikizine"; Wikipedia, "Wikipedia:Wikipedia Signpost," Wikipedia, May 7, 2009, http://en.wikipedia.org/?oldid=288444646 (accessed May 21, 2009); Wikipedia, "Wikipedia:Village Pump," Wikipedia, April 14, 2009, http://en.wikipedia.org/?oldid=283697120 (accessed July 28, 2009); for blog aggregators, see Wikimedia Foundation, "Planet Wikimedia," November 8, 2007, http://en.planet.wikimedia.org/ (accessed November 8, 2007); Millosh, "Aggregated Aggregators," millosh's blog, October 1, 2007, http://millosh.wordpress.com/2007/10/01/aggregated-aggregators/ (accessed November 8, 2007); Open, "Open Wiki Blog Planet," November 8, 2007, http://open.wikiblogplanet.com/ (accessed November 8, 2007); for podcasts, see Wikipedia, "Wikipedia:WikipediaWeekly," Wikipedia, May 30, 2008, http://en.wikipedia.org/?oldid=216003219 (accessed June 4, 2008); Wikimedia, "Wikivoices," Wikimedia, November 2008, http://meta.wikimedia.org/?oldid=1290278 (accessed December 8, 2008).

32. Wikipedia, "Wikipedia:Project Namespace," Wikipedia, November 30, 2007, http://en.wikipedia.org/?oldid=174790757 (accessed December 19, 2007); Wikimedia, "Meta," Wikimedia, October 8, 2008, http://meta.wikimedia.org/?oldid=1221501 (accessed October 31, 2008).

33. Wikipedia, "Wikipedia:Barnstars," Wikipedia, August 11, 2009, http://en.wikipedia.org/?oldid=307406807 (accessed August 20, 2009).

34. Wikipedia, "User:Raul654/Raul's Laws" (oldid=301373968)."

35. Wikipedia, "User:Malleus Fatuorum/WikiSpeak," Wikipedia, July 10, 2008, http://en.wikipedia.org/?oldid=224745874 (accessed July 10, 2008).

36. Harold Garfinkel, *Studies in Ethnomethodology* (New Jersey: Prentice-Hall, 1967), 1.

37. Alain Coulon, *Ethnomethodology*, vol. 36, Qualitative Research Methods (Thousand Oaks, CA: Sage, 1995), 29.

38. Wikipedia, "Wikipedia:Neutral Point of View," Wikipedia, November 3, 2008, http://en.wikipedia.org/?oldid=249390830 (accessed November 3, 2008).

39. Wikipedia, "Wikipedia:Attribution," Wikipedia, March 1, 2007, http://en.wikipedia.org/?oldid=111837371 (accessed March 1, 2007). Even before Wikipedia, Denis Diderot (1713–1784), editor of the *Encyclopédie,* wrote that the biases of "national prejudice" could be countered by "giving cross-references to articles where solid principles serve as the foundation for diametrically opposed truths" and that "If these cross-references, which now confirm and now refute, are carried out artistically according to a plan carefully conceived in advance, they will give the Encyclopedia what every good dictionary ought to have—the power to change men's common way of thinking." Denis Diderot, "The Encyclopedia (1755)," in *Rameau's Nephew, and Other Works*, ed. Jacques Barzun and Ralph Henry Bowen (Indianapolis: Hackett Publishing Company, 2001), 295

40. Wikipedia, "Wikipedia:No Original Research," Wikipedia, April 3, 2006, http://en.wikipedia.org/?oldid=46780169 (accessed April 6, 2006).

41. Wikipedia, "Wikipedia:Verifiability," Wikipedia, November 14, 2008, http://en.wikipedia.org/?oldid=251829388 (accessed November 14, 2008).

42. GeorgeLouis, "Wikipedia Talk:Neutral Point of View," Wikipedia, August 16, 2006, http://en.wikipedia.org/?oldid=69979085 (accessed November 14, 2007).

43. Jimmy Wales, "Wikipedia Talk:Neutral Point of View," Wikipedia, August 15, 2006, http://en.wikipedia.org/?oldid=69887768 (accessed November 14, 2007).

44. Yvonna S. Lincoln and Egon G. Guba, *Naturalistic Inquiry* (London: SAGE Publications, 1985), 39–42, 187–188, 209; Barney Glaser and Anselm Strauss, *The Discovery of Grounded Theory: Strategies for Qualitative Research* (Chicago: Aldine Publishing Company, 1967), 27, 45, 76.

45. Primary sources are managed in a mind-mapping application (Freemind). These are roughly organized into sixty top-level categories; those topics on which I spent significant time include subdivisions. So, for example, the "good faith" category includes the subcategories of compassion, humor, patience, and so on. Capturing, excerpting, annotating, and categorizing, these sources was facilitated by bibliographic tools, as described in Joseph Reagle, "Thunderdell and BusySponge," January 2009, http://reagle.org/joseph/2009/01/thunderdell.html (accessed May 28, 2009).

46. John Van Maanen, *Tales of the Field: On Writing Ethnography* (Chicago: University of Chicago Press, 1988), 1, 13.

47. Henry Jenkins, *Textual Poachers: Television Fans and Participatory Culture* (New York: Routledge, 1992), 7.

48. W. Boyd Rayward, "Visions of Xanadu: Paul Otlet (1868–1944) and Hypertext," *JASIS* 45 (1994): 235–250.

49. Wells, "The Idea of a World Encyclopedia," 921.

50. Harvey Einbinder, *The Myth of the Britannica* (New York: Grove Press, 1964); Herbert Charles Morton, *The Story of Webster's Third: Philip Gove's Controversial Dictionary and Its Critics* (New York: Cambridge University Press, 1994).

Chapter 2: The Pursuit of the Universal Encyclopedia

1. Jimmy Wales, "Hi . . ." nupedia-l, March 11, 2000, http://web.archive.org/web/20010506015648/http://www.nupedia.com/pipermail/nupedia-l/2000-March/000009.html (accessed June 7, 2006).

2. The full name is *Encyclopédie, ou Dictionnaire Raisonné des Sciences, des Arts et des Métiers* (*Encyclopedia: Or, a Systematic Dictionary of the Sciences, Arts, and Crafts*). Richard N. Schwab, "Introduction," in *Preliminary Discourse to the Encyclopedia of Diderot*, ed. Richard N. Schwab and Walter E. Rex (Indianapolis: ITT Bobbs-Merrill, 1963), xii.

3. Denis Diderot, "The Encyclopedia," in *Rameau's Nephew, and Other Works*, ed. Jacques Barzun and Ralph Henry Bowen (Indianapolis: Hackett Publishing Company, 2001), 283, 277. Essay originally published in 1755.

4. Richard R. Yeo, *Encyclopedic Visions: Scientific Dictionaries and Enlightenment Culture* (Cambridge, UK: Cambridge University Press, 2001), 57, 244. The recurrence of the idea of the social need for information "across generations" is also a theme of W. Boyd Rayward, "The Origins of Information Science and the International Institute of Bibliography/International Federation of Information and Documentation (FID)," *Journal of the American Society for Information Science* 48 (April 1997): 290.

5. Jimbo Wales, "Founder Letter Sept 2004," Wikimedia Foundation, September 2004, http://wikimediafoundation.org/wiki/Founder_letter/Founder_letter_Sept_2004 (accessed October 8, 2008).

6. Marita Sturken and Douglas Thomas, "Introduction: Technological Visions and the Rhetoric of the New," in *Technological Visions: The Hopes and Fears that Shape New Technologies*, ed. Marita Sturken, Douglas Thomas, and Sandra Ball (Philadelphia: Temple University Press, 2004), 1–18.

7. Tom Standage, *The Victorian Internet: The Remarkable Story of the Telegraph and the 19th Century's Online Pioneers* (New York: Berkley Trade, 1999), 163.

8. H. G. Wells, "World Brain: The Idea of a Permanent World Encyclopaedia," in *Encyclopédie Française* (Paris: Société des Gestion de l'Encyclopédie Française, 1937).

9. Joseph J. Corn, *The Winged Gospel: America's Romance with Aviation, 1900–1950* (New York: Oxford University Press, 1983), 129.

10. For an example of a contemporary reference to Alexandria, see Kevin Kelly, "Scan This Book! What Will Happen to Books? Reader, Take Heart! Publisher, Be Very, Very Afraid. Internet Search Engines Will Set Them Free. A Manifesto," *The New York Times Magazine* (May 14, 2006); also, see Wikimedia, "Library of Alexandria to Host Wikipedia's Annual Conference in 2008," Wikimedia Foundation, November 11, 2007, http://wikimediafoundation.org/?oldid=23186 (accessed June 12, 2008).

11. Vannevar Bush, "As We May Think," *Atlantic Monthly* (July 1945): §6, §8.

12. Wayne A. Wiegand, *A Biography of Melvil Dewey: Irrepressible Reformer* (Chicago: American Library Association, 1996); Thomas Hapke, "Wilhelm Ostwald, the 'Brücke' (Bridge), and Connections to Other Bibliographic Activities at the Beginning of the 20th Century," in *Proceedings of the 1998 Conference on the History and Heritage of Science Information Systems*, ed. M. E. Bowden, T. B. Hahn, and R. V. Williams (Information Today, Inc. for the American Society for Information Science and the Chemical Heritage Foundation, 1999), 139–147; Michael K. Buckland, *Emanuel Goldberg and His Knowledge Machine: Information, Invention, and Political Forces* (Westport, CT: Libraries Unlimited, 2006); Ronald E. Day, "Suzanne Briet: An Appreciation," *Bulletin* (January 2007).

13. Legbagede, "A Limited Company of Useful Knowledge: Paul Otlet, the International Institute of Bibliography and the Limits of Documentalism," everything2, May 18, 2001, http://everything2.com/index.pl?node_id=1053046 (accessed September 29, 2006), 1.

14. Paul Otlet, *Diaries*, quoted in W. Boyd Rayward, *The Universe of Information. The Work of Paul Otlet for Documentation and International Organisation,* 520 (Moscow: International Federation of Documentation [FID], 1975), 18.

15. Paul Otlet, "Transformations in the Bibliographical Apparatus of the Sciences," in *International Organization and Dissemination of Knowledge: Selected Essays of Paul Otlet*, ed. W. Boyd Rayward (Amsterdam: Elsevier, 1990), 148. Originally published in 1918.

16. Robert B. Goldschmidt and Paul Otlet, "On a New Form of the Book: The Microphotographic Book (1906)," in *International Organization*, see note 15 above, 93, 89.

17. Henri La Fontaine and Paul Otlet, "Creation of a Universal Bibliographic Repertory: A Preliminary Note," in *International Organization,* originally published in 1895–1896, see note 15; Otlet, "Transformations in the Bibliographical Apparatus of the Sciences," 150.

18. Rayward, "The Origins of Information Science," 290; W. Boyd Rayward, "The Case of Paul Otlet, Pioneer of Information Science, Internationalist, Visionary: Reflections on Biography," *Journal of Librarianship and Information Science* 23 (September 1991): 4.

19. "Inventions a faire" is a phrase used to describe potential technological innovations by Paul Otlet, *Traité de Documentation. Le Livre sur le Livre: Théorie et Pratique*, reprint (Liège: Centre de Lecture Publique de la Communauté Française, 1989), 389–391. Originally published in 1934. It is further discussed as "inventions to be discovered" by Rayward, "The Origins of Information Science," 18.

20. While Otlet's occasional use of superlatives and the many interrelated projects, which underwent various merges and name changes, can make it difficult, if not foolish, to speak of any single project or institution, I will limit myself to speaking of the IIB.

21. Rayward, "The Origins of Information Science," 7.

22. W. Boyd Rayward, "Visions of Xanadu: Paul Otlet (1868–1944) and Hypertext," *Journal of the American Society for Information Science* 45 (1994): 7, 14; Rayward, "The Origins of Information Science," 293.

23. I. C. McIlwaine, *Notes & Queries* (The Hague: UDC Consortium, 2006).

24. Rayward, "The Origins of Information Science," 294.

25. David Weinberger, *Everything Is Miscellaneous: The Power of the New Digital Disorder* (New York: Times Books, 2007), 92; also see Alex Wright, *Glut: Mastering Information through the Ages* (Washington, DC: John Henry Press, 2008).

26. Otlet's "monographic principle" is similar to Ostwald's "principle of the independent use of the individual piece"; for discussion of the possible origins and relations of these notions, see Hapke, "Wilhelm Ostwald, the 'Brücke' (Bridge)," 143.

27. Otlet, "Transformations in the Bibliographical Apparatus of the Sciences (1918)," 149.

28. Paul Otlet, "Something about Bibliography," in *International Organization*, see note 15, 17. Originally published in 1891–1892.

29. Bernd Frohmann, "The Role of Facts in Paul Otlet's Modernist Project of Documentation," in *European Modernism and the Information Society: Informing the Present, Understanding the Past*, ed. W. Boyd Rayward (Burlington, VT: Ashgate, 2008), 80.

30. Paul Otlet, "The Science of Bibliography and Documentation (1903)," in *International Organization*, see note 15 above, 83.

31. Ibid., 83–84.

32. Rayward, "Visions of Xanadu."

33. Alex Wright, "The Web Time Forgot," *The New York Times* (June 17, 2008).

34. Footnote 11 in W. Boyd Rayward, "H. G. Wells's Idea of a World Brain: A Critical Reassessment," *Journal of the American Society for Information Science* 50 (May 15, 1999): 32.

35. H. G. Wells, *A Modern Utopia*, 10th ed., project 6424 (Project Gutenberg EBook, 2004), §5.6. Originally published in 1905.

36. H. G. Wells, "The Open Conspiracy," December 2002, http://www.mega.nu:8080/ampp/hgwells/hg_cont.htm (accessed September 20, 2006). Originally published in 1928.

37. Dave Muddiman, "The Universal Library as Modern Utopia: The Information Society of H. G. Wells," *Library History* 14 (November 1998): 87.

38. The dystopian character of such technology use by the state is noted by Rayward, "H. G. Wells's Idea of a World Brain" and Muddiman, "The Universal Library as Modern Utopia."

39. H. G. Wells, Julian S. Huxley, and G. P. Wells, *The Science of Life: A Summary of Contemporary Knowledge about Life and Its Possibilities I, II, III* (Garden City, NY: Doubleday, Doran, 1931), 1451.

40. Otlet, "Something about Bibliography," 17; Hapke, "Wilhelm Ostwald, the 'Brücke' (Bridge)," 141.

41. H. G. Wells, "The Idea of a World Encyclopedia," *Nature* 138 (November 28, 1936): 919–921; Wells, Huxley, and Wells, *The Science of Life*, 1471; Wells, "The Idea of a World Encyclopedia," 924.

42. H. G. Wells, *World Brain* (London: Methuen, 1938), 54, 49.

43. Wells, *A Modern Utopia*, 5.

44. Wells, *World Brain*, 55.

45. Wikipedia, "Wikipedia:Neutral Point of View," Wikipedia, November 3, 2008, http://en.wikipedia.org/?oldid=249390830 (accessed November 3, 2008).

46. Wells, "The Idea of a World Encyclopedia," 921.

47. Wells's habit of working with others' work at hand is mentioned in H. G. Wells, *Experiment in Autobiography: Discoveries and Conclusions of a Very Ordinary Brain (Since 1866)* (New York: The Macmillan Company, 1934), 614–618. Accusations of plagiarism are recounted in A. B. McKillop, *The Spinster and the Prophet* (New York: Four Walls Eight Windows, 2002).

48. Suzanne Briet, *Entre Aisne et Meuse . . . et au delà*, Les cahiers ardennais 22. (Charleville-Mezières: Société de Ecrivains Ardennais, 1976), 87, quoted in Mary Niles

Maack, "The Lady and the Antelope: Suzanne Briet's Contribution to the French Documentation Movement," *Library Trends* 52, no. 4 (March 22, 2004): 5.

49. Anthony West, *H. G. Wells: Aspects of the Life* (New York: Random House, 1984), 147–148.

50. The belief in a technology-mediated peace did not completely expire, as discussed in Wendy Grossman, "Why the Net Won't Deliver World Peace," *The Daily Telegraph* (London) (December 9, 1997): 10.

51. Andrew Pam, "Xanadu FAQ," April 12, 2002, http://www.xanadu.com.au/xanadu/faq.html (accessed November 7, 2008).

52. Theodor Holm Nelson, *Literary Machines: The Report on, and of, Project Xanadu Concerning Word Processing, Electronic Publishing, Hypertext, Thinkertoys, Tomorrow's Intellectual Revolution, and Certain Other Topics Including Knowledge, Education and Freedom*, 91.1 (Sausalito, CA: Mindful Press, 1992), dedications, and beginning chapters.

53. Rayward, "Visions of Xanadu."

54. Project Xanadu, "Project Xanadu History," May 23, 2001, http://www.xanadu.com/HISTORY/ (accessed November 7, 2008).

55. Nelson, *Literary Machines*.

56. Kevin Kelly, "We Are the Web," *Wired* 13, no. 8 (August 2005).

57. Nelson, *Literary Machines*, chapter Hyperworld, 13.

58. H. G. Wells, "H. G. Wells," in *I Believe: The Personal Philosophies of Certain Eminent Men and Women of Our Time, Edited, with an Introduction and Biographical Notes, by Clifton Fadiman*, ed. Clifton Fadiman (New York: Simon & Schuster, 1939), 422–423.

59. This is similar to the observation that "Each generation imagines itself to be more intelligent than the one that went before it, and wiser than the one that comes after it" by George Orwell, *In Front of Your Nose, 1945–1950 (Collected Essays Journalism, & Letters of George Orwell)*, ed. Sonia Orwell and Ian Angus (Boston: David R. Godine, 2000).

60. Gary Wolf, "The Curse of Xanadu," *Wired* 3, no. 6 (June 1995); Theodor Holm Nelson, "Rants and Raves," *Wired* 3, no. 9 (September 1995); Theodor Holm Nelson, "Errors in 'The Curse of Xanadu,' by Gary Wolf," 1995, http://www.xanadu.com.au/ararat (accessed November 7, 2008).

61. Project Xanadu, "Project Xanadu," June 16, 2007, http://xanadu.com/ (accessed November 7, 2008); also see Theodor Holm Nelson, "I Don't Buy In," January 8, 2003, http://ted.hyperland.com/buyin.txt (accessed November 7, 2008).

62. A concise history of hypertext can be found in chapter 11 of Wright, *Glut*.

63. Michael Hart, "History and Philosophy of Project Gutenberg," Project Gutenberg, August 1992, http://www.gutenberg.org/about/history (accessed October 29, 2005).

64. Michael Hart, "Historical Draft on Digital Encyclopedias," email message to author, January 3, 2006; Robert Hobbes Zakon, "Hobbes' Internet Timeline V8.1," 2005, http://www.zakon.org/robert/internet/timeline/ (accessed January 6, 2006).

65. Hart, "History and Philosophy of Project Gutenberg."

66. Marie Lebert, *Michael Hart: Changing the World through E-Books*, English version (Project Gutenberg, 2004), http://pge.rastko.net/about/marie_lebert (accessed October 29, 2005). Originally published in French in *Edition Actu n° 90.*

67. Hart, "Historical Draft on Digital Encyclopedias."

68. Distributed Proofreaders, "Beginning Proofreaders' Frequently Asked Questions," Project Gutenberg's Distributed Proofreaders, May 27, 2004, http://www.pgdp.net/c/faq/ProoferFAQ.php (accessed October 29, 2005).

69. Yochai Benkler, "Coase's Penguin, or, Linux and the Nature of the Firm," *The Yale Law Journal* 112, no. 3 (2002): 369–446; Lee Sproull, "Online Communities," in *The Internet Encyclopedia*, ed. Hossein Bidgoli (New York: John Wiley, 2003), 733–744.

70. Nelson, *Literary Machines*, chapter Hyperworld, 7.

71. For discussion of these issues see the thread following Holden McGroin, "Encyclopaedia Britannica," December 14, 2002, http://osdir.com/ml/culture.literature.e-books.gutenberg.volunteers/2002-12/msg00016.html (accessed June 17, 2008).

72. Stephen Shankland, "Patent Reveals Google's Book-Scanning Advantage," CNet, May 4, 2009, http://news.cnet.com/8301-11386_3-10232931-76.html (accessed June 3, 2009).

73. Distributed Proofreaders, "Distributed Proofreaders Update," Distributed Proofreaders, October 19, 2005, http://www.pg-news.org/nl_archives/2004/other_2004_10_19_dp_5000_milestone.txt (accessed November 3, 2005).

74. Discussion of the incorporation of EB11 content as well as its use to identify topical blind spots in Wikipedia coverage can be found in Wikipedia, "Wikipedia:1911 Encyclopedia Topics," Wikipedia, January 20, 2008, http://en.wikipedia.org/?oldid=185703554 (accessed June 28, 2008); Wikipedia, "Wikipedia:1911 Encyclopaedia Britannica," Wikipedia, June 24, 2008, http://en.wikipedia.org/?oldid=221369034 (accessed June 28, 2008); Wikipedia, "Wikipedia Talk:1911 Encyclopaedia Britannica," Wikipedia, June 22, 2008, http://en.wikipedia.org/?oldid=221064825 (accessed June 28, 2008).

75. Hart, "Historical Draft on Digital Encyclopedias."

76. bowerbird, "The Heart and Soul of Project Gutenberg," January 30, 2004, http://osdir.com/ml/culture.literature.e-books.gutenberg.volunteers/2004-01/msg00332.html (accessed June 17, 2008).

77. Doug Wilson and Alan M. Reynard, "Interpedia Frequently Asked Questions and Answers," comp.infosystems.interpedia, January 17, 1994, http://groups.google.com/group/comp.infosystems.interpedia/msg/05d4734a3d1a03a6 (accessed October 27, 2005), Q1.

78. Wikipedia, "Interpedia," Wikipedia, May 8, 2009, http://en.wikipedia.org/?oldid=288627597 (accessed May 29, 2009).

79. WAIS was developed by Brewster Kahle, Internet entrepreneur and founder of the Internet Archive—a resource I often use to find sources that are no longer available on the Web. Throughout his career he has been an advocate for "universal access to all of human knowledge"; see Joseph Menn, "Net Archive Turns Back 10 Billion Pages of Time," dvd-discuss, October 25, 2001, http://cyber.law.harvard.edu/archive/dvd-discuss/msg15687.html (accessed March 3, 2008); Roy Rosenzweig, "Scarcity or Abundance? Preserving the Past and the Digital Era," *American Historical Review* 108, no. 3 (June 2003): 735–762.

80. Wilson and Reynard, "Interpedia Frequently Asked Questions and Answers," Q2.6.

81. Ibid.," Q1.2, Q1.1, Q4.5.

82. Ibid., Q4.2.

83. Ibid., Q3.5.2.

84. Foster Stockwell, *A History of Information Storage and Retrieval* (Jefferson, NC: Macfarlane, 2001).

85. Jorn Barger, "Beyond the Interpedia?" alt.Internet.search, August 15, 1997, http://groups.google.com/group/alt.internet.search/msg/da917af65e3ca53c?hl=en (accessed October 27, 2005).

86. Distributed Encyclopedia, "The Distributed Encyclopedia Introduction," October 8, 1999, http://web.archive.org/web/19991118223016/members.aol.com/distency/intro.htm (accessed November 3, 2005).

87. Cynthia Barnett, "Wiki Mania," *Florida Trend* (October 18, 2005).

88. Patrick Shannon, "Regarding Sanger and Shannon's Review of Y2K News Reports," January 11, 2000, http://www.greenspun.com/bboard/q-and-a-fetch-msg.tcl?msg_id=002I9H (accessed October 27, 2005).

89. Larry Sanger, "Introduction," nupedia-l, April 17, 2000, http://web.archive.org/web/20030822044803/http://www.nupedia.com/pipermail/nupedia-l/2000-April/000143.html (accessed June 7, 2006).

90. Larry Sanger, "The Early History of Nupedia and Wikipedia: A Memoir," Slashdot, April 18, 2005, http://features.slashdot.org/article.pl?sid=05/04/18/164213 (accessed April 19, 2005).

91. Liane Gouthro, "Building the World's Biggest Encyclopedia," *PC World* (March 10, 2000).

92. Larry Sanger, "I Am a Clueless Newbie," nupedia-l, March 9, 2000, http://web.archive.org/web/20030822044803/http://www.nupedia.com/pipermail/nupedia-l/2000-March/000003.html (accessed June 7, 2006).

93. Wales, "Hi . . ."

94. Larry Sanger, "602 Members," nupedia-l, March 15, 2000, http://web.archive.org/web/20030822044803/http://www.nupedia.com/pipermail/nupedia-l/2000-March/000022.html (accessed June 7, 2006).

95. Nupedia, "Nupedia.Com Editorial Policy Guidelines," November 16, 2000, http://web.archive.org/web/20010410002647/www.nupedia.com/policy.shtml (accessed October 24, 2005).

96. Sanger, "The Early History of Nupedia and Wikipedia."

97. The steps of the Nupedia process are excerpted within a case study of Wikipedia: Karim R. Lakhani and Andrew P. Mcafee, *Case Study: Wikipedia* (Boston: Harvard Business School, 2007).

98. Jimmy Wales, "Will the Real Nupedia Please Stand Up?" Slashdot, January 20, 2001, http://slashdot.org/comments.pl?sid=10074&cid=494316 (accessed October 25, 2005).

99. Richard Stallman, "Historical Draft on Digital Encyclopedias," email message to author, December 22, 2005.

100. Richard Stallman, "The Free Universal Encyclopedia and Learning Resource," Free Software Foundation, 1999, http://www.gnu.org/encyclopedia/free-encyclopedia.html (accessed October 11, 2005).

101. Erik Moeller, "Quality.Wikimedia.Org and Wikiquality-l Launched," Wikimedia, September 17, 2007, http://marc.info/?i=b80736c80709171358g6fd70693wadd14047e739a69f@mail.gmail.com (accessed September 18, 2007); Wikimedia, "Wikiquality/Portal," Wikimedia, April 14, 2009, http://meta.wikimedia.org/?oldid=1459225 (accessed June 3, 2009).

102. Stallman, "The Free Universal Encyclopedia and Learning Resource."

103. Richard Stallman, "Wale's Explaination of the Stallman Incident," wikipedia-l, October 27, 2005, http://marc.theaimsgroup.com/?l=wikipedia-l&m=113042306304247 (accessed October 27, 2005).

104. Tim Berners-Lee, *Weaving the Web: The Original Design and Ultimate Destiny of the World Wide Web by Its Inventor* (San Francisco: HarperSanFrancisco, 1999), 84.

105. Tim Berners-Lee, James Hendler, and Ora Lassila, "The Semantic Web," *Scientific American Magazine* (May 17, 2001).

106. Berners-Lee, *Weaving the Web*, 67.

107. Ibid., 70.

108. Ben Kovitz, "User:BenKovitz," Wikipedia, February 23, 2009, http://en.wikipedia.org/?oldid=272830137 (accessed February 24, 2009).

109. Larry Sanger, "Let's Make a Wiki," Nupedia, January 10, 2001, http://web.archive.org/web/20030414014355/http://www.nupedia.com/pipermail/nupedia-l/2001-January/000676.html (accessed November 15, 2005).

110. Wikipedia, "History of Wikipedia," Wikipedia, January 8, 2009, http://en.wikipedia.org/?oldid=262685944 (accessed January 8, 2009).

111. Larry Sanger, "Wikipedia Is Now Useful!" Wikipedia, June 26, 2001, http://en.wikipedia.org/wiki/Wikipedia:Announcements_2001#June_26.2C_2001 (accessed November 15, 2005).

112. Larry Sanger, "Interpedia Is Dead—Long Live Wikipedia," comp.infosystems.interpedia, September 23, 2001, http://groups.google.com/group/comp.infosystems.interpedia/msg/bb038fa078a1bf8d?hl=en (accessed November 3, 2005).

113. Wikipedia, "Nupedia," Wikipedia, December 21, 2008, http://en.wikipedia.org/?oldid=259360273 (accessed January 8, 2009).

114. Keith Hopper, "Odd Wiki CollectiveProblemSolving: Homepage," March 20, 2008, http://www.communitywiki.org/odd/CollectiveProblemSolving/HomePage (accessed March 20, 2008).

115. Sanger, "The Early History of Nupedia and Wikipedia."

116. Larry Sanger, "Re: Admins (I Suspect LMS) Permanently Deleting Things," wikipedia-l, November 4, 2001, http://marc.info/?l=wikipedia-l&m=104216623606507&w=2 (accessed November 4, 2001).

117. Jimmy Wales, "Re: Sanger's Memoirs," Wikipedia, April 20, 2005, http://marc.theaimsgroup.com/?i=4265EB8D.7020607@wikia.com (accessed November 15, 2005).

118. Citizendium, "Citizendium," Citizendium, June 17, 2008, http://en.citizendium.org/?oldid=100354482 (accessed June 30, 2008).

119. The latest cofounding controversy was prompted by an interview with Jimmy Wales when he agreed with an interviewer's question that Sanger's claims were shaky because "he basically just put himself down as co-founder" on early press

releases, in Ian Johns and Jimmy Wales, "Interview with Wikipedia Founder Jimmy Wales," Big Oak SEO Blog, April 2, 2009, http://www.bigoakinc.com/blog/interview-with-wikipedia-founder-jimmy-wales/ (accessed April 9, 2009). This prompted an "open letter" from Larry Sanger, "An Open Letter to Jimmy Wales," wikien-l, April 9, 2009, http://marc.info/?i=CDBFF612F0E04950A50B6F751478E1E9@D7WRQ591 (accessed April 9, 2009). Interestingly, the most credible and detailed account of the birth of the Wikipedia idea differs from both Wales's and Sanger's recollections; see Kovitz, "User:BenKovitz" (oldid=272830137).

120. Herb Brody, "Great Expectations: Why Technology Predictions Go Awry," in *Technology and the Future*, 7th ed., ed. Albert H. Teich (New York: St. Martin's Press, 1997), chap. 9; Paul Ceruzzi, "An Unforeseen Revolution: Computers and Expectations, 1935–1985," in *Technology and the Future*, chap. 10.

121. Sanger, "The Early History of Nupedia and Wikipedia." An example of early uncertainty about the character of Wikipedia, and the origins of the talk/discussion page, is seen in an early experience with Wikipedia: "If Wikipedia is to be an encyclopedia, then it probably is not appropriate to have threaded discussions on a subject page," by Timothy Shell, "A Few Comments About Wikipedia," wikipedia-l, January 27, 2001, http://markmail.org/message/lviwxgwjlpvrqph5 (accessed April 1, 2009).

122. Ted Pappas citing G. Donald Hudson and Walter Yust, eds., *Encyclopædia Britannica World Atlas* (Chicago: Britannica, 1956), as confirmed and identified to me by Tom Panelas, director of Britannica corporate communications, from A. J. Jacobs, *The-Know-It-All: One Man's Humble Quest to Become the Smartest Person in the World* (New York: Simon & Schuster, 2004), 341.

Chapter 3: Good Faith Collaboration

1. Wikipedia, "User:Raul654/Raul's Laws," Wikipedia, July 10, 2009, http://en.wikipedia.org/?oldid=301373968 (accessed July 20, 2009).

2. Ulrike Pfeil, Panayiotis Zaphiris, and Chee Siang Ang, "Cultural Differences in Collaborative Authoring of Wikipedia," *Journal of Computer-Mediated Communication* 12, no. 1 (2006); Andrew Lih, *The Wikipedia Revolution: How a Bunch of Nobodies Created the World's Greatest Encyclopedia* (New York: Hyperion, 2009), chap. 6.

3. Clifford Geertz, "Thick Description: Toward an Interpreted Theory of Culture," in *The Interpretation of Cultures and Local Knowledge*, ed. Clifford Geertz (New York, NY: Basic Books, 1973), chap. 1; Raymond Williams, *Keywords: A Vocabulary of Culture and Society* (New York: Oxford University Press, 1983); Marshall Sahlins, *Apologies to Thucydides: Understanding History as Culture and Vice Versa* (Chicago: The University of Chicago Press, 2004).

4. Simon Blackburn, "Culture," in *The Oxford Dictionary of Philosophy*, ed. Simon Blackburn (Oxford: Oxford University Press, 1996).

5. Edgar H. Schein, *Organizational Culture and Leadership*, 3rd ed. (San Francisco: John Wiley & Sons, 2004), 14, 32.

6. Dave Pollard, "Will That Be Coordination, Cooperation, or Collaboration?" on How to Save the World blog, March 25, 2005, http://blogs.salon.com/0002007/2005/03/25.html (accessed February 27, 2006); Patricia Montiel-Overall, "Toward a Theory of Collaboration for Teachers and Librarians," *American Library Association* 8 (2005).

7. Lore Sjberg, "The Wikipedia FAQK," April 19, 2006, http://www.wired.com/software/webservices/commentary/alttext/2006/04/70670 (accessed April 6, 2007).

8. Cass R. Sunstein, *Why Societies Need Dissent* (Cambridge, MA: Harvard University Press, 2003).

9. Steve Weber, *The Success of Open Source* (Cambridge, MA: Harvard University Press, 2004), 3.

10. For a definition and theoretical models of "interdependent decision making," see Harold H. Kelley, John G. Holmes, Norbert L. Kerr, Harry T. Reis, Caryl E. Rusbult, and Paul am Van Lange, *An Atlas of Interpersonal Situations* (New York: Cambridge, 2003).

11. Michael Schrage, *Shared Minds: The New Technologies of Collaboration* (New York: Random House, 1990), 40.

12. Henry Jenkins, *Textual Poachers: Television Fans and Participatory Culture* (New York: Routledge, 1992); Henry Jenkins, "Confronting the Challenges of Participatory Culture: Media Education for the 21st Century," Confessions of an Aca/Fan blog(October 20, 2006), http://www.henryjenkins.org/2006/10/confronting_the_challenges_of.html (accessed October 10, 2007).

13. Douglas C. Engelbart, *Augmenting Human Intellect: A Conceptual Framework* SRI Project 3578 for U.S. Air Force Office of Scientific Research, Stanford Research Institute, Menlo Park, CA, October 1962, 4D.

14. Christopher Kelty, "Geeks, Social Imaginaries, and Recursive Publics," *Cultural Anthropology* 20, no. 2 (2005): 185–214.

15. Etienne Wenger, *Communities of Practice: Learning, Meaning, and Identity* (Cambridge, UK: Cambridge University Press, 1998), 58.

16. Christian Reinhardt, "Collaborative Knowledge Creation in Virtual Communities of Practice," master's thesis, Department of Value-Process Management, Marketing (University of Innsbruck, 2003).

17. Wenger, *Communities of Practice*, 66–68.

18. Ward Cunningham, quoted in Scott Rosenberg, *Dreaming in Code: Two Dozen Programmers, Three Years, 4732 Bugs, and One Quest for Transcendent Software* (New York: Crown Publishers, 2007), 138.

19. Wikipedia, "Portland Pattern Repository," Wikipedia, June 6, 2007, http://en.wikipedia.org/?oldid=136308262 (accessed June 22, 2007).

20. Kent Beck, Mike Beedle, Arie Bennekum, Alistair Cockburn, Ward Cunningham, Martin Fowler, James Grenning, Jim Highsmith, Andrew Hunt, Ron Jeffries, Jon Kern, Brian Marick, Robert C. Martin, Steve Mellor, Ken Schwaber, Jeff Sutherland, and Dave Thomas, "Manifesto for Agile Software Development," 2001, http://agilemanifesto.org/ (accessed April 6, 2007).

21. Wikipedia, "Wikipedia Talk:Consensus," Wikipedia, July 9, 2008, http://en.wikipedia.org/?oldid=224676549 (accessed July 11, 2008).

22. Ward Cunningham, "Keynote: Wikis Then and Now," Wikimania, August 2005, http://commons.wikimedia.org/wiki/Wikimania_2005_Presentations#Keynotes (accessed August 24, 2007).

23. Wikipedia, "Wikipedia:Red Link," Wikipedia, November 15, 2007, http://en.wikipedia.org/?oldid=171608519 (accessed November 16, 2007).

24. Natalia Levina and Emmanuelle Vaast, "The Emergence of Boundary Spanning Competence in Practice: Implications for Implementation and Use of Information Systems," *MIS Quarterly* 29, no. 1 (June 1, 2005).

25. These quotes correspond to minutes 21 to 23 in Cunningham, "Keynote."

26. Nicholson Baker, "The Charms of Wikipedia," *The New York Review of Books* 55, no. 4 (March 20, 2008).

27. Wikipedia, "Wikipedia:Project Namespace," Wikipedia, November 30, 2007, http://en.wikipedia.org/?oldid=174790757 (accessed December 19, 2007); Wikipedia, "Wikipedia:Meta," Wikipedia, March 28, 2007, http://en.wikipedia.org/?oldid=118648488 (accessed April 5, 2007).

28. Meatball, "MeatballBackgrounder," Meatball Wiki, April 12, 2006, http://www.usemod.com/cgi-bin/mb.pl?MeatballBackgrounder (accessed April 12, 2006).

29. Jean Le Rond D'Alembert, *Preliminary Discourse to the Encyclopedia of Diderot (1751)*, ed. Richard N. Schwab and Walter E. Rex (Indianapolis: ITT Bobbs-Merrill, 1963), 32.

30. Susan L. Bryant, Andrea Forte, and Amy Bruckman, "Becoming Wikipedian: Transformation of Participation in a Collaborative Online Encyclopedia," in *Proceedings of the 2005 International ACM SIGGROUP Conference on Supporting Group Work* (New York: ACM, 2005).

31. Yochai Benkler, "Coase's Penguin, or, Linux and the Nature of the Firm," *The Yale Law Journal* 112, no. 3 (2002): 369–446; Lee Sproull, "Online Communities," in *The Internet Encyclopedia*, ed. Hossein Bidgoli (New York: John Wiley, 2003), 733–744.

32. Wikipedia, "Wikipedia:Be Bold," Wikipedia, August 12, 2007, http://en.wikipedia.org/?oldid=150720434 (accessed August 13, 2007).

33. Fernanda B. Viegas, Martin Wattenberg, Jesse Kriss, and Frank Ham, "Talk Before You Type: Coordination in Wikipedia," in *Proceedings of the 40th Hawaii International Conference on System Sciences* (2007); Fernanda B. Viegas, Martin Wattenberg, and Matthew M. Mckeon, "The Hidden Order of Wikipedia," in *Proceedings of HCII 2007* (2007), 445–454.

34. Stuart Geiger and David Ribes, "The Work of Sustaining Order in Wikipedia: The Banning of a Vandal," in *Proceedings of Computer Supported Collaborative Work 2010* (ACM, 2010).

35. Jimmy Wales, "Naturally Occurring Conflicts?" Air-l, March 28, 2007, http://listserv.aoir.org/pipermail/air-l-aoir.org/2007-March/012640.html (accessed March 28, 2007).

36. Jae Yun Moon and Lee Sproull, "Essence of Distributed Work: The Case of the Linux Kernel," in *Distributed Work*, ed. Pamela Hinds and Sara Kiesler (Cambridge: MIT Press, 2002), 398; Felix Stalder, "On the Differences Between Open Source and Open Culture," in *Media Mutandis: A NODE.London Reader* (2006), 4; Yochai Benkler, *The Wealth of Networks: How Social Production Transforms Markets and Freedom* (New Haven: Yale University Press, 2006), 69.

37. Bo Leuf and Ward Cunningham, *The Wiki Way: Quick Collaboration on the Web* (Boston: Addison-Wesley, 2001), 322.

38. Wikipedia, "Wikipedia:Policies and Guidelines," Wikipedia, April 10, 2007, http://en.wikipedia.org/?oldid=121662697 (accessed April 10, 2007).

39. Wikipedia, "Wikipedia Talk:Assume Good Faith," Wikipedia, March 31, 2007, http://en.wikipedia.org/?oldid=119194758 (accessed April 10, 2007).

40. Wikipedia, "Wikipedia:Policies and Guidelines" (oldid=121662697).

41. Wikipedia, "Wikipedia:Trifecta," Wikipedia, November 22, 2008, http://en.wikipedia.org/?oldid=253357789 (accessed May 29, 2009).

42. Wikipedia, "Wikipedia:Five Pillars," Wikipedia, August 15, 2009, http://en.wikipedia.org/?oldid=308208396 (accessed August 18, 2009).

43. My belief in the importance of NPOV and "good faith" is shared with Daniel H. Pink, "The Book Stops Here," *Wired* 13, no. 3 (March 2005).

44. Wikipedia, "Talk:Evolution," Wikipedia, April 6, 2007, http://en.wikipedia.org/?oldid=120622805 (accessed April 6, 2007).

45. Jimmy Wales, "Re: Sanger's Memoirs," Wikipedia, April 20, 2005, http://marc.theaimsgroup.com/?i=4265F203.3080905@wikia.com (accessed November 15, 2005);

Jimmy Wales, Fuzheado, and Liam Wyatt, "Interview W/Jimmy Wales," Wikipedia Weekly, May 8, 2008, http://wikipediaweekly.org/2008/05/08/episode-48-interview-wjimmy-wales/ (accessed June 25, 2008), minute 33.

46. Wikipedia, "User:Raul654/Raul's Laws" (oldid=301373968).

47. Wikipedia, "Troll (Internet)," Wikipedia, June 29, 2008, http://en.wikipedia.org/?oldid=222394236 (accessed July 2, 2008).

48. Wikipedia, "Wikipedia:Content Forking," Wikipedia, July 16, 2007, http://en.wikipedia.org/?oldid=144910303 (accessed August 13, 2007). This strategy of indirection and encapsulation is reminiscent of Diderot's approach to using cross-references in the *Encyclopédie*, the technique of "renvois," so as to confuse and elude censors as described in Foster Stockwell, *A History of Information Storage and Retrieval* (Jefferson, NC: Macfarlane, 2001), 91. This approach is further discussed in the contemporary context by Michael Zimmer, "Renvois of the Past, Present and Future: Hyperlinks and the Structuring of Knowledge from the Encyclopédie to Web 2.0," *New Media Society* 11, no. 95 (2009).

49. Wikipedia, "User:Salva31," Wikipedia, March 28, 2007, http://en.wikipedia.org/?oldid=118549019 (accessed April 6, 2007).

50. Wikipedia, "Talk:Evolution/Archive 2," Wikipedia, December 21, 2006, http://en.wikipedia.org/?oldid=95781264 (accessed February 7, 2007).

51. Wikipedia, "User:Graft," Wikipedia, March 23, 2007, http://en.wikipedia.org/?oldid=117319977 (accessed April 6, 2007).

52. Wikipedia, "Wikipedia:Neutral Point of View," Wikipedia, September 16, 2004, http://en.wikipedia.org/?oldid=6042007 (accessed March 5, 2004); Wikipedia, "Wikipedia:Neutral Point of View," Wikipedia, November 3, 2008, http://en.wikipedia.org/?oldid=249390830 (accessed November 3, 2008).

53. Much like my earlier comments on the ambiguity—and richness—of the concepts of culture and collaboration, "neutral" is also a provocative notion; for twenty-three different senses of the word, see Roland Barthes, *The Neutral*, trans. Rosalind E. Krauss and Denis Hollier (New York: Columbia University Press, 2005).

54. Joseph Reagle, "Is the Wikipedia Neutral?" in *Proceedings of Wikimania 2006* (2006), http://wikimania2006.wikimedia.org/wiki/Presenters/Joseph_Reagle (accessed August 1, 2006).

55. Wikipedia, "Wikipedia:Neutral Point of View," Wikipedia, April 6, 2006, http://en.wikipedia.org/?oldid=47268215 (accessed April 6, 2006).

56. Nupedia, "Nupedia.Com Editorial Policy Guidelines," November 16, 2000, http://web.archive.org/web/20010410002647/www.nupedia.com/policy.shtml (accessed October 24, 2005).

57. A. J. Jacobs, *The-Know-It-All: One Man's Humble Quest to Become the Smartest Person in the World* (New York: Simon & Schuster, 2004). A similar effort of reading the whole OED is documented by Ammon Shea, *Reading the OED: One Man, One Year, 21,730 Pages* (New York: Penguin Group, 2008).

58. Andrew Gray, "Re: Librarians, Professors, and Pundits," Wikipedia-l, July 22, 2005, http://marc.info/?i=f3fedb0d0507221318227f43c8@mail.gmail.com (accessed July 22, 2005).

59. Larry Sanger, "Epistemic Circularity," 2000, http://enlightenment.supersaturated.com/essays/text/larrysanger/diss/preamble.html (accessed October 21, 2005).

60. Objectivism WWW Service, "Objectivism-Related E-Mail Lists," December 29, 1995, http://rous.redbarn.org/objectivism/mail-lists.html (accessed November 9, 2005).

61. Larry Sanger, "The Early History of Nupedia and Wikipedia: a Memoir," Slashdot, April 18, 2005, http://features.slashdot.org/article.pl?sid=05/04/18/164213 (accessed April 19, 2005).

62. Wikipedia, "Wikipedia:Neutral Point of View" (oldid=249390830).

63. Sanger, "The Early History of Nupedia and Wikipedia."

64. Citizendium, "CZ:Neutrality Policy," Citizendium, November 25, 2007, http://en.citizendium.org/?oldid=100222609 (accessed November 8, 2008).

65. Larry Sanger, "Re: Nupedia: Questions," nupedia-l, March 10, 2000, http://web.archive.org/web/20030822044803/http://www.nupedia.com/pipermail/nupedia-l/2000-March/000006.html (accessed June 7, 2006).

66. Wikipedia, "Wikipedia:Writing for the Enemy," Wikipedia, February 8, 2009, http://en.wikipedia.org/?oldid=269389427 (accessed March 11, 2009).

67. Jimmy Wales, "Re: Bias," nupedia-l, May 21, 2000, http://web.archive.org/web/20030822044803/http://www.nupedia.com/pipermail/nupedia-l/2000-May/000343.html (accessed December 7, 2006).

68. Wikipedia, "Wikipedia:Writing for the Enemy," Wikipedia, April 18, 2006, http://en.wikipedia.org/?oldid=49041942 (accessed April 21, 2006).

69. Meatball, "AssumeGoodFaith," MeatballWiki, September 16, 2006, http://www.usemod.com/cgi-bin/mb.pl?AssumeGoodFaith (accessed February 8, 2007).

70. Wikipedia, "Wikipedia:Writing for the Enemy" (oldid=49041942).

71. Yochai Benkler and Helen Nissenbaum, "Commons-Based Peer-Production and Virtue," *The Journal of Political Philosophy* 14, no. 4 (2006): 13.

72. Leuf and Cunningham, *The Wiki Way*, 323.

73. Gabriella Coleman, "Three Ethical Moments in Debian: The Making of an (Ethical) Hacker, Part III," in *The Social Construction of Freedom in Free and Open Source Software: Actors, Ethics, and the Liberal Tradition,* (PhD thesis, University Of Chicago, 2005), chap. 6, 26.

74. Larry Wall, "Diligence, Patience, and Humility," in *Open Sources: Voices from the Open Source Revolution*, 1st ed. (Sebastopol, CA: O'Reilly, 1999).

75. Leuf and Cunningham, *The Wiki Way*, 328–329.

76. Georg von Krogh, "Care in Knowledge Creation," *California Management Review* 40, no. 3 (Spring 1998): 137.

77. Benkler and Nissenbaum, "Commons-Based Peer-Production and Virtue."

78. For possible media effects on conflict, see J. B. Walther, "Computer-Mediated Communication: Impersonal, Interpersonal, and Hyperpersonal Interaction," *Communication Research* 23 (1996): 1–43; Jay Briggs, Robert Nunamaker, Daniel Mittleman, Douglas Vogel, and Pierre Balthazard, "Lessons from a Dozen Years of Group Support Systems Research: A Discussion of Lab and Field Findings," *Journal of MIS* 13 (1997): 163–207; Raymond A. Friedman and Stephen C. Currall, "Conflict Escalation: Dispute Exacerbating Elements of E-Mail Communication," *Human Relations* 56 (2003): 1325. For conflict at the (virtual) community level, see Anna Duval Smith, "Problems of Conflict Management in Virtual Communities," in *Communities in Cyberspace*, ed. Marc Smith and Peter Kollock (London: Routledge Press, 1999), chap. 6. The author of Godwin's Law is also famous for his seminal principles of how to meet the challenges of online community; see Mike Godwin, "Nine Principles for Making Virtual Communities Work," *Wired* 2, no. 6 (June 1994).

79. Wikipedia, "Good Faith," Wikipedia, May 7, 2007, http://en.wikipedia.org/?oldid=128890878 (accessed May 17, 2007).

80. Meatball, "AssumeGoodFaith."

81. Wikipedia, "Wikipedia:Assume Good Faith," Wikipedia, April 3, 2006, http://en.wikipedia.org/?oldid=46719908 (accessed April 11, 2006).

82. For the first instance of the AGF page, see Wikipedia, "Wikipedia:Assume Good Faith," Wikipedia, March 3, 2004, http://en.wikipedia.org/?oldid=2617791 (accessed May 30, 2007); for the first comment on the talk page, see Wikipedia, "Wikipedia Talk:Assume Good Faith," Wikipedia, February 13, 2005, http://en.wikipedia.org/?oldid=10213525 (accessed May 30, 2007).

83. Wikipedia, "Wikipedia:Staying Cool When the Editing Gets Hot," Wikipedia, October 25, 2002, http://en.wikipedia.org/?oldid=383300 (accessed May 30, 2007).

84. Wikipedia, "Wikipedia:Etiquette," Wikipedia, January 24, 2004, http://en.wikipedia.org/?oldid=2219358 (accessed May 30, 2007).

85. Richard Nisbett and Lee Ross, *Human Inference* (Englewood Cliffs, NJ: Prentice Hall, 1980), 247; Daniel Kahneman and Amos Tversky, "Conflict Resolution: A Cognitive Perspective," in *Barriers to Conflict Resolution*, ed. K. Arrow et al. (New York: Norton, 1995), 47.

86. Catherine Cramton, "The Mutual Knowledge Problem and Its Consequences for Dispersed Collaboration," *Organization Science* 12 (2001): 361.

87. Wikipedia, "Hanlon's Razor," Wikipedia, January 26, 2009, http://en.wikipedia.org/?oldid=266593813 (accessed March 27, 2009).

88. Wikipedia, "Wikipedia:Assume Stupidity," Wikipedia, January 20, 2007, http://en.wikipedia.org/?oldid=102052925 (accessed May 18, 2007).

89. Meatball, "AssumeGoodFaith."

90. Wikipedia, "Wikipedia:Assume Good Faith" (oldid=46719908).

91. Wikipedia, "Wikipedia:Assume the Assumption of Good Faith," Wikipedia, February 23, 2009, http://en.wikipedia.org/?oldid=272707551 (accessed March 11, 2009).

92. Wikipedia, "Wikipedia:Assume Good Faith" (oldid=46719908).

93. Wikipedia, "Wikipedia:WikiLove," Wikipedia, January 30, 2009, http://en.wikipedia.org/?oldid=267504430 (accessed March 6, 2009).

94. Jimbo Wales, "Founder Letter Sept 2004," Wikimedia Foundation, September 2004, http://wikimediafoundation.org/wiki/Founder_letter/Founder_letter_Sept_2004 (accessed October 8, 2008).

95. Erik Moeller, "Re: Good Authors," wikien-l, June 25, 2006, http://marc.info/?i=b80736c80606251525oe0ffccdyc32f7fd9e52d6d45@mail.gmail.com (accessed June 25, 2006).

96. This is similar to the distinction between "selfish," "altruistic," and "socially concerned" classes of incentives in wiki contributions in Camille Roth, "Viable Wikis: Struggle for Life in the Wikisphere," in *WikiSym '07* (ACM, 2007).

97. A. Galinsky and T. Mussweiler, "First Offers as Anchors: The Role of Perspective Taking and Negotiator Focus," *Journal of Personality and Social Psychology* 81 (2001): 659.

98. David Bohm, *On Dialog*, ed. Lee Nichol (New York: Routledge, 1996); Daniel Yankelovich, "The Magic of Dialogue," *The Nonprofit Quarterly* 8, no. 3 (Fall 2001); Jennifer Preece and Kambiz Ghozati, "Experiencing Empathy Online," in *The Internet and Health Communication: Experience and Expectations*, ed. R. R. Rice and J. E. Katz (Thousand Oaks, CA: Sage Publications, 2001), 233; von Krogh, "Care in Knowledge Creation," 137.

99. Elinor Ostrom, "Collective Action and the Evolution of Social Norms," *The Journal of Economic Perspectives* 14, no. 3 (Summer 2000): 137–158.

100. Jennifer Preece, "Etiquette, Empathy and Trust in Communities of Practice: Stepping-Stones to Social Capital," *Journal of Universal Computer Science* 10, no. 3 (2004): 2; von Krogh, "Care in Knowledge Creation," 136; Cormac Lawler, "Wikipedia as a Learning Community: Content, Conflict and the 'Common Good'," in *Proceedings of Wikimania* (2005), http://meta.wikimedia.org/wiki/Transwiki:Wikimania 05/Paper-CL1 (accessed September 1, 2005).

101. Sirkka Jarvenpaa and Dorothy Leidner, "Communication and Trust in Global Virtual Teams," *Organization Science* 10, no. 6 (November 1999): 792.

102. Roderick M. Kramer and Peter J. Carnevale, "Trust and Intergroup Negotiation," in *Blackwell Handbook of Social Psychology: Intergroup Relations*, ed. R. Brown and S. Gaertner (Oxford, UK: Blackwell Publishers, 2001), 8.

103. Michael Sheeran, *Beyond Majority Rule: Voteless Decisions in the Religious Society of Friends* (Philadelphia, PA: Philadelphia Yearly Meeting, 1996), 66.

104. Wikipedia, "User:Raul654/Raul's Laws" (oldid=301373968).

105. For a seminal treatment of altruism, see Robert L. Trivers, "The Evolution of Reciprocal Altruism," *The Quarterly Review of Biology* 46, no. 1 (March 1971): 35–57. In a small survey of Wikipedians, there were many more "pure" or "pragmatic" altruists than "selfish" individualists, see Christian Wagner and Pattarawan Prasarnphanich, "Innovating Collaborative Content Creation: The Role of Altruism and Wiki Technology," in *Proceedings of the 40th Hawaii International Conference on System Sciences* (IEEE, 2007).

106. Wikipedia, "Wikipedia:WikiLove" (oldid=267504430).

107. Wikipedia, "Wikipedia:The World Will Not End Tomorrow," Wikipedia, March 6, 2009, http://en.wikipedia.org/?oldid=275454332 (accessed March 11, 2009).

108. Wikipedia, "Wikipedia:Ownership of Articles," Wikipedia, December 19, 2008, http://en.wikipedia.org/?oldid=259037217 (accessed December 30, 2008).

109. Jimmy Wales, "2 Questions for Jimbo about WP:OFFICE and Things That Aren't Labeled WP:OFFICE," wikien-l, April 24, 2006, http://marc.info/?i=444CF750 .7000300@wikia.com (accessed April 24, 2006).

110. Wikimedia, "Conflicting Wikipedia Philosophies," Wikimedia, 2007, http: //meta.wikimedia.org/w/index.php?title=Conflicting_Wikipedia_philosophies& (accessed May 15, 2007).

111. Wikipedia, "List of Buffy the Vampire Slayer Episodes," Wikipedia, August 6, 2009, http://en.wikipedia.org/?oldid=306433749 (accessed August 13, 2009).

112. Wikipedia, "Wikipedia:Don't Escalate," Wikipedia, February 16, 2008, http://en.wikipedia.org/?oldid=191949834 (accessed March 6, 2009).

113. Wikipedia, "Wikipedia:Candor," Wikipedia, March 18, 2009, http://en.wikipedia.org/?oldid=278065858 (accessed March 27, 2009).

114. Wikipedia, "Wikipedia:Articles for Deletion," Wikipedia, May 3, 2007, http://en.wikipedia.org/?oldid=128043238 (accessed May 15, 2007).

115. Bryan Derksen, "Re: Article Deletion (Proposed 1 Month Hiatus)," wikien-l, September 13, 2005, http://marc.info/?i=432663FA.20806@shaw.ca (accessed September 13, 2005).

116. Wikipedia, "Wikipedia:Please Do Not Bite the Newcomers," Wikipedia, October 16, 2008, http://en.wikipedia.org/?oldid=245559838 (accessed November 3, 2008).

117. Wikipedia, "Wikipedia:Do Not Disrupt Wikipedia to Illustrate a Point," Wikipedia, March 5, 2009, http://en.wikipedia.org/?oldid=275119939 (accessed March 6, 2009).

118. Wikipedia, "Wikipedia:Wikipedia Is Not Therapy," Wikipedia, December 26, 2008, http://en.wikipedia.org/?oldid=260174442 (accessed May 29, 2009).

119. Karl Fogel, "Producing Open Source Software: How to Run a Successful Free Software Project," 2005, http://producingoss.com/en/producingoss.html (accessed February 4, 2008), 49.

120. Jimmy Wales, "Re: To: Jimmy Wales—Admin-Driven Death of Wikipedia," wikien-l, June 4, 2006, http://marc.info/?i=44833A94.3080606@wikia.com (accessed June 4, 2006).

121. Sanger, "The Early History of Nupedia and Wikipedia."

122. Wikipedia, "Wikipedia:Policies and Guidelines" (oldid=121662697). Wikis are said to be based on the concept that "people really can be polite and well mannered," in Leuf and Cunningham, *The Wiki Way*, 323.

123. Wikipedia, "Wikipedia:Civility," Wikipedia, May 28, 2006, http://en.wikipedia.org/?oldid=55584755 (accessed May 28, 2006).

124. Mark Kingwell, *A Civil Tongue: Justice, Dialogue and the Politics of Pluralism* (University Park, PA: Penn State University Press, 1995), 247.

125. Wikipedia, "Wikipedia:Civility," Wikipedia, March 5, 2009, http://en.wikipedia.org/?oldid=275263680 (accessed March 6, 2009).

126. Kingwell, *A Civil Tongue*, 249.

127. Wikipedia, "Wikipedia:Civility" (oldid=55584755).

128. Wikipedia, "User:Raul654/Raul's Laws" (oldid=301373968), law 250.

129. Steven Shapin, *A Social History of Truth: Civility and the Science in Seventeenth-Century England* (Chicago: The University Of Chicago Press, 1994).

130. Analysis of discourse within online communities finds that politeness is correlated with technical topics and with being a Wikipedia administrator; see Moira Burke and Robert Kraut, "Mind Your Ps and Qs: The Impact of Politeness and Rudeness in Online Communities," in *Proceedings of the ACM 2008 Conference on Computer Supported Cooperative Work* (New York: ACM, 2008), 37–46; Moira Burke and Robert Kraut, "Mopping Up: Modeling Wikipedia Promotion Decisions," in ibid., 37–46.

131. Wikipedia, "Wikipedia:Civility/Poll," Wikipedia, August 12, 2009, https://secure.wikimedia.org/wikipedia/en/wiki/Wikipedia:Civility/Poll (accessed August 12, 2009).

132. Wikipedia, "Wikipedia:Department of Fun," Wikipedia, March 15, 2009, http://en.wikipedia.org/?oldid=277324016 (accessed March 18, 2009).

133. The Onion, "Congress Abandons WikiConstitution," *The Onion* 41, no. 39 (September 28, 2005); The Onion, "Wikipedia Celebrates 750 Years of American Independence," *The Onion* 42, no. 30 (July 28, 2006).

134. Wikipedia, "Image:Penny Arcade Comic-20051216H.Jpg," Wikipedia, May 3, 2007, http://en.wikipedia.org/?oldid=128059883 (accessed May 18, 2007).

135. Wikipedia, "Category:Wikipedia Humor," Wikipedia, March 2009, http://en.wikipedia.org/?oldid=274733123 (accessed March 18, 2009).

136. Wikimedia, "Category:Humor Songs and Poems," Wikimedia, 2006, http://meta.wikimedia.org/?oldid=386342 (accessed October 23, 2006).

137. Daniel R. Tobias, "I Am the Very Model of a Modern Wikipedian," wikien-l, October 8, 2006, http://marc.info/?i=45282D0D.27600.830BAC1@dan.tobias.name (accessed October 8, 2006).

138. Wikipedia, "Wikipedia:Assume Bad Faith," Wikipedia, April 30, 2007, http://en.wikipedia.org/?oldid=127274333 (accessed May 18, 2007).

139. Wikipedia, "Wikipedia:Neutral Point of View" (oldid=6042007).

140. Wikimedia, "Don't Be Dense," Wikimedia, March 27, 2006, http://meta.wikimedia.org/?oldid=314371 (accessed April 12, 2006).

141. Wikimedia, "AWWDMBJAWGCAWAIFDSPBATDMTD," March 4, 2008, http://meta.wikimedia.org/?oldid=903231 (accessed May 6, 2008).

142. Wikipedia, "User:Raul654/Raul's Laws" (oldid=301373968).

143. Schrage, *Shared Minds*, 164.

144. Wikipedia, "Wikipedia:Sarcasm Is Really Helpful," Wikipedia, February 20, 2009, http://en.wikipedia.org/?oldid=272003931 (accessed March 11, 2009).

145. Death Phoenix, "April Fool's Day Proposal," wikien-l, March 6, 2006, http://marc.info/?i=5c5f03dd0603061414q106fb34eyfda2976db69fb39b@mail.gmail.com (accessed March 6, 2006); Wikipedia, "Wikipedia:Pranking," Wikipedia, January 7, 2009, http://en.wikipedia.org/?oldid=262416467 (accessed March 18, 2009).

146. Christopher Thieme, "Incivility, Re: Psychosis," wikien-l, January 15, 2007, http://marc.info/?i=b9bebe7a0701151206n42ebdbccse91c90af62c6ea98@mail.gmail.com (accessed January 15, 2007).

147. Wikipedia, "Wikipedia:WikiLove" (oldid=267504430).

148. Leuf and Cunningham, *The Wiki Way*, 323.

149. Wikipedia, "Wikipedia:Collaboration First," Wikipedia, February 5, 2009, http://en.wikipedia.org/?oldid=268796710 (accessed March 6, 2009).

Chapter 4: The Puzzle of Openness

1. Wikipedia, "Wikipedia, the Free Encyclopedia," Wikipedia, November 14, 2008, http://en.wikipedia.org/?oldid=251850832 (accessed December 2, 2008).

2. Douglas Rushkoff, *Open Source Democracy: How Online Communication Is Changing Offline Politics* (London: Demos, 2003); Eric Krangel, "Two Attempts at Opening Up Religion Online," NewAssignment.Net, January 9, 2007, http://www.newassignment.net/blog/eric_krangel/jan2007/01/religion_gets_op (accessed January 11, 2007); I review much of this usage in Joseph Reagle, "Notions of Openness," in *FM10 Openness: Code, Science, and Content: Selected Papers from the First Monday Conference*, (First Monday, 2006), http://reagle.org/joseph/2006/02/fm10-openness (accessed December 20, 2007).

3. Robert Michels, *Political Parties*, trans. Eden Paul and Cedar Paul (Ontario: Batoche Books, 2001), §6.2, originally published in 1911; Jo Freeman, *The Tyranny of Structurelessness: Why Organisations Need Some Structure to Ensure They Are Democratic)* (Struggle, 1996), 1, originally published in 1970; Mitch Kapor, quoted in Joseph Reagle, "Internet Quotation Appendix," Working Draft, Berkman Center for Internet and Society, Harvard Law School, Cambridge, MA, March 1999.

4. Wikipedia, "Wikipedia:Ignore All Rules," Wikipedia, July 1, 2008, http://en.wikipedia.org/?oldid=222901357 (accessed July 2, 2008).

5. Wikipedia, "Wikipedia:What 'Ignore All Rules' Means," Wikipedia, September 17, 2007, http://en.wikipedia.org/?oldid=158433756 (accessed September 17, 2007).

6. Jimbo Wales, "Statement of Principles," Wikipedia, October 27, 2001, http://en.wikipedia.org/?oldid=89244123 (accessed March 20, 2007).

7. Joseph Reagle, "Open Content Communities," *M/C: A Journal of Media and Culture* 7, no. 3 (July 2004).

8. Robert Neelly Bellah, Richard Madsen, William M. Sullivan, Ann Swidler, and Steven M. Tipton, *Habits of the Heart: Individualism and Commitment in American Life* (Berkeley: University Of California Press, 1996), 333.

9. Samuel Bowles and Herbert Gintis, "The Moral Economy of Communities: Structured Populations and the Evolution of Pro-Social Norms," *Evolution & Human Behavior* 19, no. 1 (1998): 3–25; Lee Sproull, Caryn A. Conley, and Jae Yun Moon, "Prosocial Behavior on the Net," in *The Social Net: The Social Psychology of the Internet*, ed. Yair Amichahi-Hamburger (Oxford: Oxford University Press, 2004), chap. 6, 139–162.

10. Reagle, "Open Content Communities." See also Joseph Reagle, "Open Communities and Closed Law," in *In the Shade of the Commons—Toward a Culture of Open Networks*, ed. Lipika Bansal, Paul Keller, and Geert Lovink (Amsterdam: Waag Society, 2006), 165–167.

11. Richard M. Stallman, "The Free Software Definition," GNU, 2005, http://www.gnu.org/philosophy/free-sw.html (accessed April 5, 2006); OSI, "The Open Source Definition," OSI, July 24, 2006, http://www.opensource.org/docs/definition.php (accessed November 7, 2007); Freedom Defined, "Definition of Free Cultural Works," Freedom Defined, December 1, 2008, http://freedomdefined.org/Definition (accessed July 7, 2009).

12. See Chris Dibona, Sam Ockman, and Mark Stone, *Open Sources: Voices from the Open Source Revolution* (Sebastopol, CA: O'Reilly, 1999); Sam Williams, *Free as in Freedom: Richard Stallman's Crusade for Free Software* (Sebastopol, CA: O'Reilly, 2002); Steve Weber, *The Success of Open Source* (Cambridge: Harvard University Press, 2004); Samir Chopra and Scott Dexter, *Decoding Liberation: the Promise of Free and Open Source Software* (New York: Routledge, 2007).

13. Williams, *Free as in Freedom*.

14. Richard M. Stallman, "The GNU Project," GNU, June 16, 2005, http://www.gnu.org/gnu/thegnuproject.html (accessed April 5, 2006).

15. OSI, "History of the OSI," OSI, September 19, 2006, http://www.opensource.org/history (accessed November 7, 2007).

16. Jimmy Wales quoted in Stacy Schiff, "Know It All: Can Wikipedia Conquer Expertise?" *New Yorker* (July 31, 2006): 3; referring to Eric Raymond, "The Cathedral and the Bazaar," 1997, http://www.catb.org/~esr/writings/cathedral-bazaar/ (accessed April 29, 2004).

17. Wales, "Statement of Principles" (oldid=89244123).

18. Eric Raymond, quoted in Schiff, "Know It All," 8.

19. dharma, "Agreed," Kuro5hin, December 30, 2004, http://www.kuro5hin.org/story/2004/12/30/142458/25 (accessed December 31, 2004).

20. Felix Stalder, "On the Differences between Open Source and Open Culture," *NODE.London, March* 2006, http://publication.nodel.org/On-the-Differences (accessed June 5, 2006).

21. Lawrence Lessig, *Free Culture: How Big Media Uses Technology and the Law to Lock Down Culture and Control Creativity* (New York: The Penguin Press, 2004); a history of Creative Commons and the free culture movement is presented by David Bollier, *Viral Spiral: How the Commoners Built a Digital Republic of Their Own* (New York: the new press title case, 2009).

22. Meatball, "MeatballBackgrounder," Meatball Wiki, April 12, 2006, http://www.usemod.com/cgi-bin/mb.pl?MeatballBackgrounder (accessed April 12, 2006); Meatball, "FairProcess," Meatball Wiki, November 9, 2007, http://www.usemod.com/cgi-bin/mb.pl?FairProcess (accessed November 9, 2007).

23. On computers eroding accountability, see Helen Nissenbaum, "Accountability in a Computerized Society," *Science and Engineering Ethics* 2, no. 2 (1996): 25–42; on furthering accountability, see Shay David, "Opening the Source of Accountability," *First Monday* 9, no. 11 (November 2004); Beth Simone Noveck, "Wiki-Government: How Open-Source Technology Can Make Government Decision-Making More Expert and More Democratic," *Democracy: A Journal of Ideas*, no. 7 (Winter 2007).

24. Jill Coffin, "Analysis of Open Source Principles and Diverse Collaborative Communities," *First Monday* 11, no. 6 (May 2006).

25. Sean Hansen, Nicholas Berente, and Kalle Lyytinen, "Wikipedia, Critical Social Theory, and the Possibility of Rational Discourse," *The Information Society* 25, no. 1 (January 2009): 38–59.

26. Wikimedia, "Steward Policies," Wikimedia, October 25, 2007, http://meta.wikimedia.org/?oldid=724037 (accessed November 2, 2007).

27. Wales, "Statement of Principles" (oldid=89244123); Wikipedia, "Wikipedia:Please Do Not Bite the Newcomers," Wikipedia, March 5, 2009, http://en.wikipedia.org/?oldid=275109391 (accessed March 6, 2009).

28. Tony Sidaway, "Wikipedia and Autism," wikien-l, October 11, 2005, http://marc.info/?i=605709b90510110104w6bcdf9d6g7f1fe5945d0d05a6@mail.gmail.com (accessed October 11, 2005).

29. Wales, "Statement of Principles" (oldid=89244123).

30. Jimmy Wales, "Re: A Proposal for the New Software," wikipedia-l, October 18, 2001, http://marc.info/?l=wikipedia-l&m=104216623606384&w=2 (accessed March 7, 2007).

31. Jimmy Wales, "Why I Oppose a Cabal," Nostalgia Wikipedia, October 26, 2001, http://nostalgia.wikipedia.org/?oldid=24911 (accessed March 7, 2008).

32. Ronline, "The Wikipedia Ombudsman," wikien-l, January 7, 2006, http://marc.info/?i=648f108b0601070134l616eb900u4f97b75522540e0f@mail.gmail.com (accessed January 7, 2006); Kelly Martin, "An Example of Transparency," Nonbovine Ruminations, June 19, 2007, http://nonbovine-ruminations.blogspot.com/2007/06/example-of-transparency.html (accessed June 19, 2007).

33. Wikipedia, "User:Raul654/Raul's Laws," Wikipedia, July 10, 2009, http://en.wikipedia.org/?oldid=301373968 (accessed July 20, 2009).

34. Meatball, "RightToFork," Meatball Wiki, October 11, 2007, http://www.usemod.com/cgi-bin/mb.pl?RightToFork (accessed October 11, 2007).

35. Weber, *The Success of Open Source*, 159, 92.

36. David A. Wheeler, "Why Open Source Software/Free Software (OSS/FS)? Look at the Numbers!" May 9, 2005, http://www.dwheeler.com/oss_fs_why.html (accessed October 1, 2004).

37. Ascander Suarez and Juan Ruiz, "The Spanish Fork of Wikipedia," in *Proceedings of Wikimania* (2005) http://meta.wikimedia.org/wiki/Transwiki:Wikimania05/Paper-AS1 (accessed September 1, 2005).

38. Larry Sanger, "Toward a New Compendium of Knowledge," Citizendium, September 15, 2006, http://citizendium.org/essay.html (accessed December 20, 2007).

39. Larry Sanger, "How's the Unforking Going?" Citizendium Forums, January 2007, http://forum.citizendium.org/index.php?topic=459.30;wap2 (accessed November 9, 2007); Larry Sanger, "Our Gift to the World: CC-by-Sa," Citizendium Blog, December 21, 2007, http://blog.citizendium.org/2007/12/21/our-gift-to-the-world-cc-by-sa/ (accessed March 20, 2008). In November 2008 the GNU Foundation and Wikimedia Foundation did make it easier for content to flow between Wikimedia and Creative Commons projects, including Citizendium, by making the GFDL and CC licenses compatible/interoperable; see Mike Linksvayer, "Wikipedia/CC News: FSF Releases FDL 1.3," Creative Commons Blog, November 3, 2008, http://creativecommons.org/weblog/entry/10443 (accessed November 3, 2008).

40. Jimmy Wales, "Bias and Open Content Licensing," nupedia-l, May 8, 2000, http://web.archive.org/web/20030822044803/http://www.nupedia.com/pipermail/nupedia-l/2000-May/000222.html (accessed June 7, 2006).

41. Wikipedia, "Wikipedia, the Free Encyclopedia" (oldid=251850832).

42. Wikipedia, "Wikipedia:What Wikipedia Is Not," Wikipedia, May 28, 2009, http://en.wikipedia.org/?oldid=292975573 (accessed May 29, 2009).

43. Because the term *anonymous* is persistently used by the community, despite my caveat that it really means "not logged in," I will hereafter use it myself without further qualification unless required by the context.

44. Evan Lehmann, "Rewriting History under the Dome," *Lowell Sun Online* (January 27, 2006); Wikipedia, "Congressional Staffer Edits to Wikipedia," Wikipedia, February 18, 2007, http://en.wikipedia.org/?oldid=109060581 (accessed March 8, 2007).

45. Wikipedia, "Help:CheckUser," Wikipedia, June 29, 2007, http://en.wikipedia.org/?oldid=141475772 (accessed November 9, 2007).

46. Wikipedia, "Wikipedia:Banning Policy," Wikipedia, March 16, 2007, http://en.wikipedia.org/?oldid=115595742 (accessed March 16, 2007).

47. Wikipedia, "Wikipedia:What Wikipedia Is Not" (oldid=292975573).

48. John Seigenthaler, "A False Wikipedia 'Biography,'" *USA Today* (November 29, 2005).

49. Associated Press, "Author of False Wikipedia Entry Apologizes: Tennessee Man Says False Entry on Seigenthaler Part of Joke on Co-Worker," MSNBC, December 12, 2005, http://www.msnbc.msn.com/id/10439120/ (accessed December 12, 2005).

50. Jimmy Wales, "Wikipedia and Defamation: Man Apologizes after Fake Wikipedia Post," Air-l, December 13, 2005, http://listserv.aoir.org/pipermail/air-l-aoir.org/2005-December/008894.html (accessed December 14, 2005).

51. Wikipedia, "Wikipedia:Blocking Policy Proposal," Wikipedia, November 5, 2006, http://en.wikipedia.org/?oldid=85910798 (accessed March 16, 2007).

52. Katie Hafner, "Growing Wikipedia Refines Its 'Anyone Can Edit' Policy," *The New York Times* (June 17, 2006).

53. Jimmy Wales, "Re: New York Times Article," wikien-l, June 20, 2006, http://marc.info/?i=44987D15.2070408@wikia.com (accessed June 20, 2006).

54. Hafner, "Growing Wikipedia Refines Its 'Anyone Can Edit' Policy."

55. Wikipedia, "Wikipedia:Blocking Policy," Wikipedia, November 26, 2008, http://en.wikipedia.org/?oldid=254316018 (accessed November 27, 2008).

56. Fl Celloguy, "Re: Blocking Proposal," wikien-l, October 18, 2005, http://marc.info/?i=BAY114-F237B1A8334738B323C83E4C7710@phx.gbl (accessed October 18, 2005).

57. Anthony DiPierro, "Re: Blocking Proposal," wikien-l, October 19, 2005, http://marc.info/?i=71cd4dd90510191024r4332fe14tbb075711698d60c0@mail.gmail.com (accessed October 19, 2005).

58. Wikipedia, "Wikipedia:Blocking Policy Proposal" (oldid=85910798).

59. Clay Shirky, *Here Comes Everybody: The Power of Organizing without Organizations* (New York: Penguin Press, 2007), 253.

60. For a philosophical understanding of the types of values and how they are often in conflict, see Thomas Nagel, "The Fragmentation of Value," in *Mortal Questions* (Cambridge, UK: Cambridge University Press, 1979), 128–141. For more recent work on identifying and balancing values in technical design, see Mary Flanagan, Daniel Howe, and Helen Nissenbaum, "Embodying Values in Technology: Theory and Practice," in *Information Technology and Moral Philosophy*, ed. Jeroen van den Hoven and John Weckert (Cambridge, UK: Cambridge University Press, 2006), chap. 6, 9.

61. Langdon Winner, "Do Artifacts Have Politics?" in *The Whale and the Reactor* (Chicago: University of Chicago Press, 1986), 18–39.

62. Jimmy Wales, quoted in Hafner, "Growing Wikipedia Refines Its 'Anyone Can Edit' Policy."

63. Steve Woolgar, "The Turn to Technology in Social Studies of Science," *Science, Technology, & Human Values* 16, no. 1 (Winter 1991): 20–50; Batya Friedman and Helen Nissenbaum, "Bias in Computer Systems." *ACM Transactions in Information Systems* 14, no. 2 (1996): 330–346; Helen Nissenbaum, "How Computer Systems Embody Values," *Computer* 34, no. 3 (March 2001): 120–118.

64. Jake Wartenberg and Ragesoss, "Flagged Revisions," Wikipedia, January 24, 2009, http://en.wikipedia.org/?oldid=266468243 (accessed January 26, 2009); Noam Cohen, "Wikipedia May Restrict Public's Ability to Change Entries," NYTimes.com (January 23, 2009), http://bits.blogs.nytimes.com/2009/01/23/wikipedia-may-restrict-publics-ability-to-change-entries/?apage=2 (accessed January 26, 2009).

65. Wikipedia, "Wikipedia:Five Pillars," Wikipedia, August 15, 2009, http://en.wikipedia.org/?oldid=308208396 (accessed August 18, 2009).

66. Wales, "Statement of Principles" (oldid=89244123).

67. Seigenthaler, "A False Wikipedia 'Biography.'"

68. Jimmy Wales, "Re: 'Should Not Be Written by an Interested Party,'" wikien-l, May 3, 2006, http://marc.info/?i=4458EF10.9060705@wikia.com (accessed May 3, 2006).

69. Erik Moeller, "Indefinite Block and Desysopping by User:Danny," wikien-l, April 19, 2006, http://marc.info/?i=b80736c80604191156y3ba64470vab33c6fb0d513a1d@mail.gmail.com (accessed April 19, 2006).

70. Wikipedia, "Wikipedia:Oversight," Wikipedia, July 12, 2009, http://en.wikipedia.org/?oldid=301752049 (accessed July 16, 2009).

71. Wikipedia, "Wikipedia:Wikipedia Signpost/2007-01-15/2006 in Review," Wikipedia, January 19, 2007, http://en.wikipedia.org/?oldid=101894518 (accessed January 23, 2007).

72. Larry Sanger, "Why Collaborative Free Works Should Be Protected by the Law," Slashdot, December 7, 2005, http://slashdot.org/~LarrySanger/journal/123625 (accessed December 7, 2005).

73. Jimmy Wales, quoted in Richard Pérez-Peña, "Keeping News of Kidnapping off Wikipedia," *The New York Times* (June 28, 2009).

74. Anonymous, "David S. Rohde (Difference Between Revisions)," Wikipedia, June 20, 2009, http://en.wikipedia.org/?oldid=297562979&diff=297567037 (accessed July 16, 2009).

75. Max Weber, *Economy and Society: An Outline of Interpretive Sociology*, ed. Claus Wittich Guenther Roth, vol. 1 (Berkeley: University of California Press, 1978), 212–302, originally published in 1914; Wolfgang J. Mommsen, *The Political and Social Theory of Max Weber: Collected Essays* (Chicago: University of Chicago Press, 1992), 42.

76. Clay Shirky, "Wikis, Grafitti, and Process," Many-to-Many blog, August 1, 2003, http://many.corante.com/20030801.shtml#50187 (accessed August 22, 2006).

77. Wikipedia, "Wikipedia:Practical Process," Wikipedia, June 24, 2008, http://en.wikipedia.org/?oldid=221497230 (accessed July 11, 2008).

78. Wikipedia, "Wikipedia:Practical Process," Wikipedia, September 20, 2006, http://en.wikipedia.org/?oldid=76785518 (accessed September 20, 2006).

79. Wikipedia, "Wikipedia:Avoid Instruction Creep," Wikipedia, March 26, 2007, http://en.wikipedia.org/?oldid=118096776 (accessed March 27, 2007); Ivan Beschastnikh, Travis Kriplean, and David W. Mcdonald, "Wikipedian Self-Governance in Action: Motivating the Policy Lens," in *ICWSM 2008: International Conference on Weblogs and Social Media* (2008), 4.

80. Andrew Lih, "Unwanted: New Articles in Wikipedia," August 23, 2007, http://www.andrewlih.com/blog/2007/07/10/unwanted-new-articles-in-wikipedia/ (accessed August 27, 2007).

81. Brian Butler, Elisabeth Joyce, and Jacqueline Pike, "Don't Look Now, but We've Created a Bureaucracy: The Nature and Roles of Policies and Rules in Wikipedia," in *CHI '08: Proceeding of the Twenty-Sixth Annual SIGCHI Conference on Human Factors in Computing Systems, ed. Eytan Adar et al. (Menlo Park, CA: The AAAI Press, 2008), 8.*

82. Wikipedia, "Wikipedia:WikiLawyering," Wikipedia, March 7, 2007, http://en.wikipedia.org/?oldid=113216995 (accessed March 15, 2007); Ryan McGrady, "Gaming Against the Greater Good," *First Monday* 14, no. 2 (February 2, 2009).

83. Phil Sandifer, "I've Kicked the Process Habit," wikien-l, September 16, 2006, http://marc.info/?i=A9F21CA7-852B-4684-AA83-0F759F8DDFBA@gmail.com (accessed September 16, 2006).

84. Gary A. Yukl, *Leadership in Organizations*, 1st ed. (Englewood Cliffs, NJ: Prentice-Hall, 1981), 18.

85. For discussion of cabals, see Bryan Pfaffenberger, "If I Want It, It's Okay: Usenet and the (Outer) Limits of Free Speech," *The Information Society* (1996): 365–386; for use of the *dictator* term, see Wikipedia, "Benevolent Dictator for Life," Wikipedia, May 11, 2009, http://en.wikipedia.org/?oldid=289287807 (accessed May 29, 2009).

86. Walter van Kalken, "Re: Secret Admin List," wikipedia-l, May 19, 2006, http://marc.info/?i=446E5989.9070807@vankalken.net (accessed May 19, 2006).

87. Cass R. Sunstein, *Why Societies Need Dissent* (Cambridge, MA: Harvard University Press, 2003), 158.

88. For the announcement, see Angela, "Introducing a New Mailing List," wikien-l, December 4, 2006, http://marc.info/?i=8b722b800612032144j1711060by4a60db0e8a051460@mail.gmail.com (accessed December 4, 2006); for the description, see WikiChix, *WikiChix* (WikiChix, 2007).

89. Alphax, "Re: Fwd: Request to Mailing List Wikichix-l Rejected," wikien-l, December 5, 2006, http://marc.info/?i=45758194.7020503@gmail.com (accessed December 5, 2006).

90. Guettarda, "Re: Fwd: Request to Mailing List Wikichix-l Rejected," wikien-l, December 5, 2006, http://marc.info/?i=47683e960612050629t45577488xdf97707a966de6cb@mail.gmail.com (accessed December 5, 2006).

91. Yuwei Lin, "Inclusion, Diversity and Gender Equality: Gender Dimensions in the Free/Libre Open Source Software Development," in *Gender and IT* (IDEA Groups, 2005).

92. Fiona Wilson, "Can Compute, Won't Compute: Women's Participation in the Culture of Computing," *New Technology, Work and Employment* 18, no. 2 (2003): 127–142.

93. Bogdan Giusca, "Re: Fwd: Request to Mailing List Wikichix-l Rejected," wikien-l, December 5, 2006, http://marc.info/?i=281363007.20061205165218@dapyx.com (accessed December 5, 2006).

94. Robert Cawdrey, *A Table Alphabetical of Hard Usual English Words*, ed. Ian Lancashire (Web Development Group University of Toronto Library, 1997). Originally published in 1604/1966.

95. Foster Stockwell, *A History of Information Storage and Retrieval* (Jefferson, NC: Macfarlane, 2001), 111; Tom McArthur, *Worlds of Reference: Lexicography, Learning, and Language from the Clay Tablet to the Computer* (Cambridge, UK: Cambridge University Press, 1986), 107; a replication of these plates is provided in Herman Kogan, *The Great EB: The Story of the Encyclopaedia Britannica* (Chicago: University Of Chicago Press, 1958).

96. Wayne A. Wiegand, *A Biography of Melvil Dewey: Irrepressible Reformer* (Chicago: American Library Association, 1996).

97. Gillian Thomas, *A Position to Command Respect: Women and the Eleventh Britannica* (Metuchen, NJ: The Scarecrow Press, 1992).

98. Oded Nov, "What Motivates Wikipedians?" *Communications of the ACM* 50, no. 11 (November 2007); Ruediger Glott, Philipp Schmidt, and Rishab Ghosh, "Wikipedia Survey—First Results," April 9, 2009, http://upload.wikimedia.org/wikipedia/foundation/a/a7/Wikipedia_General_Survey-Overview_0.3.9.pdf (accessed April 24, 2009).

99. Guettarda, "Re: Fwd: Request to Mailing List Wikichix-l Rejected," wikien-l, December 5, 2006, http://marc.info/?i=47683e960612050657s6c91283cr8458565ce1a776b6@mail.gmail.com (accessed December 5, 2006).

100. Gabriella Coleman, "Three Ethical Moments in Debian: The Making of an (Ethical) Hacker, Part III," in *The Social Construction of Freedom in Free and Open Source Software: Actors, Ethics, and the Liberal Tradition* (PhD thesis, University Of Chicago, 2005), chap. 6.

101. Jimmy Wales, "Re: Nazi Userboxes," wikien-l, December 1, 2006, http://marc.info/?i=4570B0FF.6030209@wikia.com (accessed December 1, 2006).

102. Wikipedia, "User:The Cunctator," Wikipedia, May 18, 2009, http://en.wikipedia.org/?oldid=290621049 (accessed May 29, 2009).

103. Wikipedia, "Wikipedia:Do Not Disrupt Wikipedia to Illustrate a Point," Wikipedia, March 5, 2009, http://en.wikipedia.org/?oldid=275119939 (accessed March 6, 2009).

104. Jennifer Vesperman and Deb Richardson, "Frequently Asked Questions," January 4, 2002, http://www.linuxchix.org/about-linuxchix.html (accessed January 23, 2007).

105. Erinn Clark, "Frequently Asked Questions," Debian Women, 2005, http://women.debian.org/faqs/ (accessed January 23, 2007).

106. Ubuntu-Women Project, "Ubuntu Women," Ubuntu-Women Project, 2006, http://ubuntu-women.org/ (accessed January 23, 2007).

107. Webmaster, "KDE Women Homepage—Contact," KDE, January 23, 2007, http://women.kde.org/contact/ (accessed January 23, 2007).

108. Bryan Derksen, "Re: Fwd: Request to Mailing List Wikichix-l Rejected," wikien-l, December 5, 2006, http://marc.info/?i=4575ABC6.50801@shaw.ca (accessed December 5, 2006).

109. Freeman, *The Tyranny of Structurelessness.*

110. Sunstein, *Why Societies Need Dissent,* 158.

111. Nancy Fraser, "Rethinking the Public Sphere: A Contribution to the Critique of Actually Existing Democracy," in *Habermas and the Public Sphere, ed. Craig Calhoun* (Cambridge: The MIT Press, 1992), chap 5, 123.

112. Sunstein, *Why Societies Need Dissent,* 160–161.

113. Clay Shirky, "News of Wikipedia's Death Greatly Exaggerated," Many-to-Many, May 25, 2006, http://many.corante.com/archives/2006/05/25/news_of_wikipedias_death_greatly_exaggerated.php (accessed May 25, 2006).

Chapter 5: The Challenges of Consensus

1. H. G. Wells, "The Idea of a World Encyclopedia," *Nature* 138 (November 28, 1936): 921.

2. Wikipedia, "Creation-Evolution Controversy," Wikipedia, July 27, 2008, http://en.wikipedia.org/?oldid=228137973 (accessed August 7, 2008).

3. Wikipedia, "Wikipedia:Working Group on Ethnic and Cultural Edit Wars," Wikipedia, January 23, 2008, http://en.wikipedia.org/?oldid=186309462 (accessed January 23, 2008).

4. On the organization of knowledge, and alphabetization in particular, see Peter Burke, *A Social History of Knowledge: From Gutenberg to Diderot* (Cambridge, UK: Polity Press, 2000), 109–115; Daniel Headrick, *When Information Came of Age* (Oxford: Oxford University Press, 2000), 154, 162–163; Foster Stockwell, *A History of Information Storage and Retrieval* (Jefferson, NC: Macfarlane, 2001), chap. 11.

5. The new mode of dealing with information is described by David Weinberger, *Everything Is Miscellaneous: The Power of the New Digital Disorder* (New York: Times Books, 2007).

6. Wikipedia, "Wikipedia:Disambiguation," Wikipedia, July 17, 2008, http://en.wikipedia.org/?oldid=226312626 (accessed July 18, 2008).

7. Wikipedia, "Buffy," Wikipedia, July 12, 2008, http://en.wikipedia.org/?oldid=225134221 (accessed July 18, 2008).

8. Wikipedia, "Wikipedia:Consensus," Wikipedia, June 1, 2009, http://en.wikipedia.org/?oldid=293735821 (accessed June 10, 2009).

9. Wikipedia, "Wikipedia:Arbitration Committee," Wikipedia, July 16, 2008, http://en.wikipedia.org/?oldid=225940072 (accessed July 21, 2008).

10. For discussion of the extent to which the ArbCom interprets or makes policy see John Lee, "Re: BADSITES ArbCom Case About to Close," wikien-l, October 17, 2007, http://marc.info/?i=6fb0c9b00710170801p7e54742csd2dd12ebb6c71930@mail

.gmail.com (accessed October 17, 2007); Risker, "Re: ArbCom Legislation," wikien-l, June 18, 2008, http://marc.info/?i=eb45e7c0806180917v5b217175pc469da2ab683a699@mail.gmail.com (accessed June 18, 2008).

11. "Evidence" is often densely sourced hyperlinks to various discussions and specific Wikipedia edits; I remove such references (e.g., "[32]") from excerpts as they serve no purpose here.

12. Wikipedia, "Wikipedia:Requests for Arbitration/Naming Conventions," Wikipedia, January 20, 2007, http://en.wikipedia.org/?oldid=101952465 (accessed July 16, 2008).

13. Ibid.

14. Wiktionary, "Consensus," July 11, 2008, http://en.wiktionary.org/?oldid=4617397 (accessed July 11, 2008); Lawrence E. Susskind, "A Short Guide to Consensus Building," in *Consensus Building Handbook*, ed. Lawrence E. Susskind, Sarah McKearnen, and Jennifer Thomas-Lamar (Thousand Oaks, CA: Sage, 1999), part 1, 600.

15. Wikipedia, "Consensus Decision-Making," Wikipedia, July 10, 2008, http://en.wikipedia.org/?oldid=224735826 (accessed July 11, 2008).

16. Sheeran is affirming nine characteristics first articulated by Stuart Chase, *Roads to Agreement: How to Get Along Better with Other People.* (New York: Harper & Brothers, 1951), 51–52; Michael Sheeran, *Beyond Majority Rule: Voteless Decisions in the Religious Society of Friends* (Philadelphia, PA: Philadelphia Yearly Meeting, 1996), 51.

17. Andrea Forte and Amy Bruckman, "Scaling Consensus: Increasing Decentralization in Wikipedia Governance," in *Proceedings of the Proceedings of the 41st Annual Hawaii International Conference on System Sciences* (Washington, DC: IEEE Computer Society, 2008).

18. Lawrence E. Susskind, Sarah McKearnen, and Jennifer Thomas-Lamar, *Consensus Building Handbook* (Thousand Oaks, CA: Sage, 1999).

19. Tim Berners-Lee, *Weaving the Web: The Original Design and Ultimate Destiny of the World Wide Web by Its Inventor* (San Francisco: HarperSanFrancisco, 1999), 62–63.

20. A history of the IETF, including its own sometimes troubled relationship with the traditional International Organization for Standardization (ISO) can be found in Andrew L. Russell, "'Rough Consensus and Running Code' and the Internet-OSI Standards War," *IEEE Annals of the History of Computing* 28, no. 3 (July 2006): 48–61; a history of the Web and W3C can be found in Berners-Lee, *Weaving the Web.*

21. Ian Jacobs, "World Wide Web Consortium Process Document," W3C, October 14, 2005, http://www.w3.org/2005/10/Process-20051014/policies.html#Consensus (accessed October 14, 2005).

22. The W3C's—and Berners-Lee's—influence is questioned in Simson Garfinkel, "The Web's Unelected Government," *Technology Review* (November 1998). Also, Ber-

ners-Lee's role was challenged when those interested in developing and "monetizing" Web services objected to his "controversial push to develop a type of artificial intelligence called the Semantic Web," Paul Festa, "Critics Clamor for Web Services Standards," CNET News.com, July 10, 2002, http://news.cnet.com/2102-1023_3-834990.html (accessed September 2, 2007).

23. David D. Clark, "A Cloudy Crystal Ball: Visions of the Future," IETF, July 1992, http://xys.ccert.edu.cn/reference/future_ietf_92.pdf (accessed March 26, 1999), 19.

24. Scott Bradner, "RFC 2418: IETF Working Group Guidelines and Procedures," IETF, September 1998, http://tools.ietf.org/html/rfc2418 (accessed March 26, 1999).

25. Wikipedia, "Wikipedia:Consensus," Wikipedia, July 5, 2008, http://en.wikipedia.org/?oldid=223708449 (accessed July 11, 2008).

26. Bradner, "RFC 2418," § 3.3.

27. Jacobs, "World Wide Web Consortium Process Document," § 3.3.

28. Wikipedia, "Wikipedia:Consensus" (oldid=223708449).

29. Delirium, "Re: Writing Style (Was: a Valid Criticism)," wikien-l, October 7, 2005, http://marc.info/?i=4346DE10.8030902@hackish.org (accessed October 7, 2005).

30. Wikipedia, "Wikipedia:Consensus" (oldid=223708449).

31. Wikimedia, "Foundation Issues," Wikimedia, March 26, 2006, http://meta.wikimedia.org/?oldid=312949 (accessed April 6, 2006); Wikipedia, "Wikipedia:Consensus" (oldid=223708449). For mutability of consensus, see Travis Kriplean, Ivan Beschastnikh, David W. Mcdonald, and Scott A. Golder, "Community, Consensus, Coercion, Control: CS*W or How Policy Mediates Mass Participation," in *GROUP'07: International Conference on Supporting Group Work 2007* (New York: ACM, 2007), 173.

32. Tim Berners-Lee, "Hypertext Style: Cool URIs Don't Change," W3C, 1998, http://www.w3.org/Provider/Style/URI (accessed July 31, 2008).

33. Rowan Collins, "Re: Multilingual Redirect," Wikipedia-l, October 21, 2005, http://marc.info/?i=9f02ca4c0510211419u6e4f9d04m@mail.gmail.com (accessed October 21, 2005).

34. Wikipedia, "Wikipedia:Consensus," Wikipedia, August 23, 2007, http://en.wikipedia.org/?oldid=153141886 (accessed July 16, 2008).

35. Philip Sandifer, "Reflections on the End of the Spoiler Wars," wikien-l, November 14, 2007, http://marc.info/?i=9D7A631E-A3E7-4083-A91C-E3BB9C848704@gmail.com (accessed November 14, 2007).

36. Bradner, "RFC 2418," § 3.3.

37. Ibid., § 3.3.

38. Wikipedia, "Wikipedia:Consensus" (oldid=223708449).

39. Gerard Meijssen, "Re: New Request for Cantonese Wikipedia: Vote at 29-6," Wikipedia-l, September 25, 2005, http://marc.info/?i=43370ACD.3030301@gmail.com (accessed September 25, 2005).

40. Jane J. Mansbridge, *Beyond Adversary Democracy* (New York: Basic Books, 1980), 32.

41. Bradner, "RFC 2418," §3.3.

42. Wikipedia, "User:Raul654/Raul's Laws," Wikipedia, July 10, 2009, http://en.wikipedia.org/?oldid=301373968 (accessed July 20, 2009).

43. Sheeran, *Beyond Majority Rule*, 92–98.

44. Wikipedia, "Wikipedia:Requests for Arbitration/Naming Conventions" (oldid=101952465).

45. Ibid.

46. My selective portrayal of this case should not be construed as representative of the case or the participants.

47. Wikipedia, "Wikipedia Talk:Requests for Arbitration/Naming Conventions," Wikipedia, February 12, 2007, http://en.wikipedia.org/?oldid=107562939 (accessed July 16, 2008).

48. Wikipedia, "Wikipedia:Requests for Arbitration/Naming Conventions" (oldid=101952465).

49. Wikipedia, "Wikipedia Talk:Requests for Arbitration/Naming Conventions" (oldid=107562939).

50. Wikipedia, "Wikipedia:Voting Is Evil," Wikipedia, August 6, 2007, http://en.wikipedia.org/?oldid=149531708 (accessed November 1, 2007). Voting is supposed to be exceptional enough that Wikipedians keep a tally of the number of times a hundred people or more—now two hundred—have voted on something, as seen in Wikipedia, "Wikipedia:Times That 100 Wikipedians Supported Something," Wikipedia, May 29, 2009, http://en.wikipedia.org/?oldid=292998287 (accessed May 29, 2009).

51. Wikipedia, "Wikipedia:Straw Polls," Wikipedia, May 23, 2008, http://en.wikipedia.org/?oldid=214504581 (accessed October 3, 2008).

52. Wikimedia, "Don't Vote on Everything," Wikimedia, 2007, http://meta.wikimedia.org/?oldid=646278 (accessed November 1, 2007).

53. Wikimedia, "Polls Are Evil," Wikimedia, 2007, http://meta.wikimedia.org/?oldid=656194 (accessed November 1, 2007); Wikipedia, "Wikipedia:Polling Is

Not a Substitute for Discussion," Wikipedia, September 10, 2008, http://en.wikipedia.org/?oldid=237594693 (accessed October 3, 2008).

54. Mansbridge, *Beyond Adversary Democracy*, 175.

55. Meatball, "VotingIsGood," January 27, 2005, http://www.usemod.com/cgi-bin/mb.pl?VotingIsGood (accessed July 11, 2008). Meatball still uses the historic "CamelCase" wiki method of linking to other pages by concatenating the capitalized words of the article name.

56. Clay Shirky, "A Group Is Its Own Worst Enemy," Keynote at O'Reilly Emerging Technology Conference, shirky.com, April 24, 2003, http://www.shirky.com/writings/group_enemy.html (accessed June 27, 2007), 14.

57. Malcolm Gladwell, *The Tipping Point: How Little Things Can Make a Big Difference* (Boston: Little Brown, 2000), 181.

58. Wikipedia, "Wikipedia:Consensus" (oldid=223708449).

59. Cass R. Sunstein, *Why Societies Need Dissent* (Cambridge, MA: Harvard University Press, 2003), 59.

60. Meatball, "VotingIsEvil," November 16, 2007, http://www.usemod.com/cgi-bin/mb.pl?VotingIsEvil (accessed July 11, 2008).

61. Wikipedia, "Wikipedia:Sock Puppetry," Wikipedia, March 5, 2009, http://en.wikipedia.org/?oldid=275114814 (accessed March 6, 2009).

62. Ben McIlwain, "We Need a Policy Against Vote-Stacking," wikien-l, May 4, 2006, http://marc.info/?i=44594B8C.1030001@gmail.com (accessed May 4, 2006); Wikipedia, "Wikipedia:Articles for Deletion," Wikipedia, May 3, 2007, http://en.wikipedia.org/?oldid=128043238 (accessed May 15, 2007); Wikipedia, "Wikipedia:Requests for Adminship," Wikipedia, August 6, 2008, http://en.wikipedia.org/?oldid=230207377 (accessed August 6, 2008).

63. Danny Wool, "[Foundation-l] Breaking Promises (Was Re: Where We Are Headed)," Foundation-l, June 5, 2006, http://lists.wikimedia.org/pipermail/foundation-l/2006-June/020815.html (accessed August 8, 2006).

64. Mansbridge, *Beyond Adversary Democracy*, 10.

65. Ryan Paul, "XML Spec Editor: OOXML ISO Process Is 'Unadulterated BS,'" March 2, 2008, http://arstechnica.com/news.ars/post/20080302-xml-spec-editor-ooxml-iso-process-is-unadulterated-bs.html (accessed August 6, 2008).

66. Edward Macnaghten, "ODF/OOXML Technical White Paper," May 2, 2007, http://www.freesoftwaremagazine.com/articles/odf_ooxml_technical_white_paper (accessed August 6, 2008).

67. Scott M. Fulton III, "Evidence of Microsoft Influencing OOXML Votes in Nordic States," BetaNews, August 28, 2007, http://www.betanews.com/article/Evidence_of

_Microsoft_Influencing_OOXML_Votes_in_Nordic_States/1188335569 (accessed August 6, 2008).

68. Martin Bryan, "Report on WG1 Activity for December 2007 Meeting of ISO/IEC JTC1/SC34/WG1 in Kyoto," November 29, 2007, http://www.jtc1sc34.org/repository/0940.htm (accessed August 5, 2008).

69. I discuss similar arguments over trading votes and advocacy that took place in the KDE development community when their bug-tracking system adopted a feature allowing users to vote for their most pressing bugs in Joseph Reagle, "Bug Tracking Systems as Public Spheres," *Techné: Research in Philosophy and Technology* 11, no. 1 (Fall 2007).

70. John Tex, "We Need to Recognize That Advocating Is a Basic Right," wikien-l, May 4, 2006, http://marc.info/?i=eeec757b0605040857m212c525818165c43f970228ab@mail.gmail.com (accessed May 4, 2006).

71. Tony Sidaway, "Re: We Need to Recognize That Advocating Is a Basic Right," wikien-l, May 5, 2006, http://marc.info/?i=605709b90605050526g75eb04b4o17ebdba61acf3899@mail.gmail.com (accessed May 5, 2006).

72. Zoney, "Re: 'Consensus' and Decision Making on Wikipedia," wikien-l, June 29, 2007, http://marc.info/?l=wikien-l&m=118311257100564&w=2 (accessed June 29, 2007).

73. Tony Sidaway, "Re: 'Consensus' and Decision Making on Wikipedia," wikien-l, June 29, 2007, http://marc.info/?l=wikien-l&m=118312000423989&w=2 (accessed June 29, 2007).

74. Tony Sidaway, "Re: 'Consensus'and Decision Making on Wikipedia," wikien-l, June 27, 2007, http://marc.info/?l=wikien-l&m=118292148913045&w=2 (accessed June 27, 2007).

75. Adrian, "Re: 'Consensus' and Decision Making on Wikipedia," wikien-l, June 26, 2007, http://marc.info/?i=46818E71.2050000@googlemail.com (accessed June 26, 2007).

76. Marc Riddell, "Re: 'Consensus' and Decision Making on Wikipedia," wikien-l, June 26, 2007, http://marc.info/?i=C2A70AEC.5E08%michaeldavid86@comcast.net (accessed June 26, 2007).

77. Bibliomaniac15, "How Many Wikipedians Does It Take to Screw in a Lightbulb?" Wikipedia, July 11, 2009, http://en.wikipedia.org/?oldid=301565313 (accessed August 20, 2009); Lise Broer, "How Many Wikipedians Does It Take to Change a Light Bulb?" Durova, August 24, 2008, http://durova.blogspot.com/2008/08/how-many-wikipedians-does-it-take-to.html (accessed August 20, 2009).

78. Wikipedia, "User:Malleus Fatuorum/WikiSpeak," Wikipedia, July 10, 2008, http://en.wikipedia.org/?oldid=224745874 (accessed July 10, 2008).

Chapter 6: The Benevolent Dictator

1. Amitai Etzioni, *Comparative Analysis of Complex Organizations* (NY: Free Press of Glencoe, 1975).

2. Jimmy Wales, "Re: Neo-Nazis to Attack Wikipedia," wikien-l, February 7, 2005, http://marc.info/?i=20050207162911.GO29080@wikia.com (accessed February 7, 2005).

3. Joseph Reagle, "Do as I Do: Authorial Leadership in Wikipedia," in *WikiSym '07: Proceedings of the 2007 International Symposium on Wikis* (New York: ACM Press, 2007); a similar model to the "authorial" mode of leadership is offered by Mathieu O'Neil, *Cyberchiefs: Autonomy and Authority in Online Tribes* (New York: Pluto Press, 2009).

4. For emergent leadership, see Bernard M. Bass, *Bass & Stogdill's Handbook of Leadership: Theory, Research, and Managerial Applications*, 3rd ed. (New York: Free Press, 1990), 126–127. For characteristics of emergent leadership in the online context, see Gianluca Bosco, "Implicit Theories of 'Good Leadership' in the Open-Source Community" (master's thesis, Technical University of Denmark, 2004); Youngjin Yoo and Maryam Alavi, "Emergent Leadership in Virtual Teams: What Do Emergent Leaders Do?" *Information and Organization* 14, no. 1 (January 2004): 27–58. The role of technical contributions in such communities is discussed by David Waguespack and Lee Fleming, "Penguins, Camels, and Other Birds of a Feather: Brokerage, Boundary Spanning, and Leadership in Open Innovation Communities," April 8, 2005, http://opensource.mit.edu/papers/flemingwaguespack.pdf (accessed January 13, 2006).

5. For the mixing of governance models in software development, see O'Neil, *Cyberchiefs*; Gabriella Coleman, "Three Ethical Moments in Debian: The Making of an (Ethical) Hacker, Part III," in *The Social Construction of Freedom in Free and Open Source Software: Actors, Ethics, and the Liberal Tradition* (PhD thesis, University Of Chicago, 2005), chap. 6; Siobhan O'Mahony and Fabrizio Ferraro, "Governance in Production Communities," April 2007, http://www.business.ualberta.ca/tcc/documents/TII_3_OMahoney_Ferraro_final.pdf (accessed June 15, 2007). Similarly, the larger Internet culture, a "set of values and beliefs informing behavior," is argued by to be constituted by technomeritocratic, hacker, virtual communitarian, and entrepreneurial values by Manuel Castells, *The Internet Galaxy: Reflections on the Internet, Business, and Society* (Oxford: Oxford University Press, 2001), 36–37. Theories of governance are compared in the Wikipedia context in Christopher Goldspink, "Social Self-Regulation in Computer Mediated Communities: the Case of Wikipedia," *International Journal of Agent Technologies and Systems* 1, no. 1 (2009): 19–33.

6. Jimmy Wales, "From Jimbo Wales' user talk page," quoted in Wikimedia, "Meta:Talk:Benevolent Dictator," Wikimedia, March 16, 2007, http://meta.wikimedia.org/?oldid=544462 (accessed May 21, 2008).

7. "Charismatic authority" is seminally discussed by Max Weber, *Economy and Society: An Outline of Interpretive Sociology*, vol. 1, ed. Claus Wittich Guenther Roth (Berkeley: University of California Press, 1978), 215, originally published in 1914; in the online context, see Giampaolo Garzarelli and Roberto Galoppini, "Capability Coordination in Modular Organization: Voluntary FS/OSS Production and the Case of Debian GNU/Linux," in Economics Working Paper Archive at WUST, no. 0312005 (Industrial Organization, 2003), 18; O'Neil, *Cyberchiefs*, 18.

8. Eric Raymond, "Homesteading the Noosphere," *First Monday* 3, no. 10 (1998): 15.

9. Ibid.; Wikipedia, "Benevolent Dictator For Life," Wikipedia, May 11, 2009, http://en.wikipedia.org/?oldid=289287807 (accessed May 29, 2009).

10. Edwin P. Hollander, "Competence and Conformity in the Acceptance of Influence," *Journal of Abnormal and Social Psychology* 61, no. 3 (1960): 365–369; Gary A. Yukl, *Leadership in Organizations*, 1st ed. (Englewood Cliffs, NJ: Prentice-Hall, 1981), 29; Raymond, "Homesteading the Noosphere," 15.

11. Meatball, "GodKing," Meatball Wiki, November 2, 2007, http://www.usemod.com/cgi-bin/mb.pl?GodKing (accessed November 2, 2007).

12. David A. Wheeler, "Why Open Source Software/Free Software (OSS/FS)? Look at the Numbers!" May 9, 2005, http://www.dwheeler.com/oss_fs_why.html (accessed October 1, 2004).

13. For examples of similar anxiety and humor about leadership, see Bryan Pfaffenberger, "If I Want It, It's Okay: Usenet and the (Outer) Limits of Free Speech," *The Information Society* 12, no. 4 (1996): 379; Wikipedia, "Wikipedia:Mediation Cabal," Wikipedia, February 16, 2009, http://en.wikipedia.org/?oldid=271167724 (accessed May 29, 2009), 379.

14. Edgar H. Schein, *Organizational Culture and Leadership*, 2nd ed. (San Francisco: Jossey-Bass Publishers, 1992), 231.

15. Larry Sanger, "The Early History of Nupedia and Wikipedia: A Memoir," Slashdot, April 18, 2005, http://features.slashdot.org/article.pl?sid=05/04/18/164213 (accessed April 19, 2005).

16. Jimmy Wales as quoted in Marshall Poe, "The Hive: Can Thousands of Wikipedians Be Wrong? How an Attempt to Build an Online Encyclopedia Touched Off History's Biggest Experiment in Collaborative Knowledge," *The Atlantic Monthly* (September 2006): 2.

17. Sanger, "The Early History of Nupedia and Wikipedia."

18. Anthere, "Re: Sanger's Memoirs," wikipedia-l, April 23, 2005, http://marc.info/?i=4269F0E4.8080501@yahoo.com (accessed April 23, 2005).

19. Evan Prodromou, *Your Paper and Presentation at WikiSym*, email message to author, October 23, 2007.

20. Stacy Schiff, "Know It All: Can Wikipedia Conquer Expertise?" *The New Yorker* (July 31, 2006): 3; Jimmy Wales, "Citing Material from the Web?" nupedia-l, October 10, 2000, http://web.archive.org/web/20030822044803/http://www.nupedia.com/pipermail/nupedia-l/2000-October/000617.html (accessed June 7, 2006).

21. Steve Bennett, "[OT] Jimbo Wales' Edit Count (Was: Re: Here's an," wikien-l, February 1, 2006, http://marc.info/?i=f1c3529e0602011213peed6b47idd8aa61c56053fd2@mail.gmail.com (accessed February 1, 2006).

22. For example, between June and September 2008 Wales did not post to wikiEN-l MARC, "Viewing Messages Posted by 'Jimmy Wales,'" MARC, May 23, 2008, http://marc.info/?a=111340136500001&r=1&w=2 (accessed October 3, 2008); his edits can be seen at Wikipedia, "Jimbo Wales User Contributions," Wikipedia, May 23, 2008, http://en.wikipedia.org/wiki/Special:Contributions/Jimbo_Wales (accessed October 3, 2008).

23. Schein, *Organizational Culture and Leadership*, 231.

24. Jimmy Wales, "Re: Do We NEED an Article on Feminism for the One on Masculism?" nupedia-l, September 15, 2000, http://web.archive.org/web/20030822044803/http://www.nupedia.com/pipermail/nupedia-l/2000-September/000572.html (accessed June 7, 2006).

25. Jimmy Wales, "Re: A Neo-Nazi Wikipedia," wikien-l, August 23, 2005, http://marc.info/?i=430B6975.9040906@wikia.com (accessed August 23, 2005).

26. Jimmy Wales, "Re: Cruft," wikien-l, September 11, 2005, http://marc.info/?i=43244CE3.1060905@wikia.com (accessed September 11, 2005).

27. Jimbo Wales, "User Talk:Jimbo Wales," March 3, 2007, http://en.wikipedia.org/?oldid=112270687 (accessed March 3, 2007).

28. Jimmy Wales, "Re: Just What *Is* Jimbo's Role Anyway?" wikien-l, March 21, 2007, http://marc.info/?i=4600EC9C.3020909@wikia.com (accessed March 21, 2007).

29. Jimmy Wales, "Wikimania 2007 Interview with Joseph Reagle," August 3, 2007.

30. Jimmy Wales, "Re: Ant: Serbo-Croatian Wikipedia — a Policy Question?" wikipedia-l, October 16, 2005, http://marc.info/?i=43524BFD.9000501@wikia.com (accessed October 16, 2005).

31. Jimmy Wales, "Re: Request for New Wikipedia: Friulian," wikipedia-l, November 4, 2004, http://marc.info/?i=20041104204518.GE24033@wikia.com (accessed November 4, 2004).

32. Jimmy Wales, "Re: Status of Wikimedia," wikipedia-l, November 7, 2005, http://marc.info/?i=436FDD60.1060301@wikia.com (accessed November 7, 2005).

33. Jimmy Wales, "Re: Re: Static HTML Dumps," wikipedia-l, October 18, 2005, http://marc.info/?l=wikipedia-l&m=112961483513485&w=2 (accessed October 18, 2005).

34. Mark Williamson, "Re: Wikimedia Foundation Internal Radio System," wikipedia-l, April 12, 2005, http://marc.info/?i=849f98ed050412011473cf4b91@mail.gmail.com (accessed April 12, 2005).

35. Some of the more extreme message boards have come to popular attention through media coverage of "cyberbullying," see Mattathias Schwartz, "Malwebolence—The World of Web Trolling," NYTimes.com, August 3, 2008, http://www.nytimes.com/2008/08/03/magazine/03trolls-t.html (accessed August 14, 2008).

36. Jimmy Wales, "Wikimedia Board Elections," wikipedia-l, September 16, 2006, http://marc.info/?l=wikipedia-l&m=115839007632183&w=2 (accessed September 16, 2006); Wikimedia Foundation, "Wikimedia Foundation Announces Board Elections," Wikimedia, June 27, 2007, http://wikimediafoundation.org/wiki/Board_Elections_2007 (accessed June 27, 2007).

37. Jake Wartenberg and Ragesoss, "Flagged Revisions," Wikipedia, January 24, 2009, http://en.wikipedia.org/?oldid=266468243 (accessed January 26, 2009); Noam Cohen, "Wikipedia May Restrict Public's Ability to Change Entries," NYTimes.com, January 23, 2009, http://bits.blogs.nytimes.com/2009/01/23/wikipedia-may-restrict-publics-ability-to-change-entries/ (accessed January 26, 2009); Jimmy Wales, "User Talk:Jimbo Wales," Wikipedia, January 22, 2009, http://en.wikipedia.org/?oldid=266489946 (accessed January 26, 2009).

38. Wikipedia, "Wikipedia:Administrators," March 23, 2005, http://en.wikipedia.org/?oldid=11508036 (accessed May 4, 2007).

39. Wikipedia, "Wikipedia:Practical Process," Wikipedia, June 24, 2008, http://en.wikipedia.org/?oldid=221497230 (accessed July 11, 2008); Wikipedia, "Wikipedia:Lists of Protected Pages," Wikipedia, June 30, 2008, http://en.wikipedia.org/?oldid=222655315 (accessed October 2, 2008).

40. Wikipedia, "Wikipedia:Administrators" (oldid=11508036).

41. Wikipedia, "Wheel (Unix Term)," Wikipedia, September 13, 2008, http://en.wikipedia.org/?oldid=238091451 (accessed September 30, 2008).

42. Wikipedia, "Wikipedia:Administrators," Wikipedia, October 2, 2008, http://en.wikipedia.org/?oldid=242442563 (accessed October 2, 2008).

43. Wikipedia, "Wikipedia:List of Administrators," Wikipedia, August 11, 2009, http://en.wikipedia.org/?oldid=307325114 (accessed August 11, 2009); Wikipedia, "Wikipedia:Bureaucrats," Wikipedia, July 16, 2009, http://en.wikipedia.org/?oldid=302392524 (accessed July 17, 2009).

44. Wikimedia, "Stewards," Wikimedia, July 9, 2009, http://meta.wikimedia.org /?oldid=1547805 (accessed July 17, 2009).

45. The structure and character of Wikipedia governance is further explored by Andrea Forte and Amy Bruckman, "Scaling Consensus: Increasing Decentralization in Wikipedia Governance," in *Proceedings of the Proceedings of the 41st Annual Hawaii International Conference on System Sciences* (Washington, DC: IEEE Computer Society, 2008).

46. Wikipedia, "Wikimedia Foundation," Wikipedia, July 14, 2009, http://en.wiki pedia.org/?oldid=302106222 (accessed July 17, 2009).

47. Erik Moeller, "Rename Admins to Janitors," wikien-l, March 6, 2007, http: //marc.info/?i=b80736c80703060517k56dfabadg2e9f9bf3720cb3a9@mail.gmail.com (accessed March 6, 2007).

48. Wikimedia, "Stewards," Wikimedia, November 2, 2007, http://meta.wikimedia .org/?oldid=733931 (accessed November 2, 2007).

49. Wikimedia, "Steward Policies," Wikimedia, October 25, 2007, http://meta.wiki media.org/?oldid=724037 (accessed November 2, 2007).

50. Wikimedia Foundation, "Board of Trustees/Restructure Announcement," Wiki media Foundation, April 26, 2008, http://wikimediafoundation.org/wiki/Board_of _Trustees/Restructure_Announcement (accessed June 11, 2009).

51. Wikimedia Foundation, "Board of Trustees," October 13, 2005, http://wikimedi afoundation.org/?oldid=10702 (accessed August 26, 2008).

52. Erik Moeller, "RK Temp-Banned," wikien-l, October 2, 2003, http://marc .info/?l=wikien-l&m=106513613512602&w=2 (accessed April 16, 2009); Jimmy Wales, "Wikiquette 'Committee,'" wikien-l, October 2, 2003, http://marc .info/?l=wikien-l&m=106513629912729&w=2 (accessed April 16, 2009); Wikipedia, "Wikipedia:Arbitration Committee," Wikipedia, July 16, 2008, http://en.wikipedia .org/?oldid=225940072 (accessed July 21, 2008).

53. Wikipedia, "User:Raul654/Raul's Laws," Wikipedia, July 10, 2009, http://en.wiki pedia.org/?oldid=301373968 (accessed July 20, 2009).

54. Wikimedia, "Polls Are Evil," Wikimedia, 2007, http://meta.wikimedia.org /?oldid=656194 (accessed November 1, 2007).

55. Wikipedia, "Wikipedia:Deletion Guidelines for Administrators," Wikipedia, July 4, 2008, http://en.wikipedia.org/?oldid=223498575 (accessed October 3, 2008).

56. Wikipedia, "Wikipedia:New Admin School," Wikipedia, June 20, 2008, http: //en.wikipedia.org/?oldid=220558698 (accessed July 31, 2008); Wikipedia, "Wikipedia:Admin Coaching," Wikipedia, July 24, 2008, http://en.wikipedia .org/?oldid=227653331 (accessed July 31, 2008).

57. Wikipedia, "Wikipedia:Advice for New Administrators," Wikipedia, June 10, 2008, http://en.wikipedia.org/?oldid=218417159 (accessed August 8, 2008).

58. Pfaffenberger, "If I Want It, It's Okay." Coleman, "Three Ethical Moments in Debian."

59. Raymond, "Homesteading the Noosphere," 11.

60. For the earliest instance I could find in computer communities see Steve Dyer, "Re on the Direction of Net.Motss," net.motss, October 2, 1984, http://groups.google.com/group/net.motss/msg/216bcdcffc0aadfe (accessed May 2, 2007); for application to Torvalds, see Russell Nelson, "Patchlevel 6," comp.os.linux, September 22, 1992, http://groups.google.com/group/comp.os.linux/msg/1a0a82ff3c662036 (accessed May 2, 2007; Charles Hedrick, "386Bsd Vs Linux," comp.os.linux, October 3, 1992, http://groups.google.com/group/comp.os.linux/msg/cebaafdf43eebec7 (accessed May 2, 2007).

61. NSK, "Re: Flags," wikipedia-l, January 10, 2005, http://marc.info/?i=200501101748.06696.nsk2@wikinerds.org (accessed January 10, 2005).

62. Karl Fogel, "Producing Open Source Software: How to Run a Successful Free Software Project," 2005, http://producingoss.com/en/producingoss.html (accessed February 4, 2008), 48.

63. Geoff Cohen, quoted in Clay Shirky, "A Group Is Its Own Worst Enemy," Keynote at O'Reilly Emerging Technology Conference, shirky.com, April 24 2003, http://www.shirky.com/writings/group_enemy.html (accessed June 27, 2007), 5.

64. SlimVirgin, "Re: Original Research versus Point of View," wikien-l, January 18, 2005, http://marc.info/?i=4cc603b0501181558569cb84f@mail.gmail.com (accessed January 18, 2005).

65. James D. Forrester, "Re: Your Golorous [Sic] Leader." wikien-l, April 8, 2005, http://marc.info/?i=200504081652.j38GqMtd009896@mail-relay-2.csv.warwick.ac.uk (accessed April 8, 2005).

66. Yukl, *Leadership in Organizations*, 144.

67. Anthere, "Re: NYTimes.Com: Google May Host Encyclopedia Project," wikipedia-l, February 14, 2005, http://marc.info/?i=4211339B.6030803@yahoo.com (accessed February 14, 2005).

68. Wikipedia, "Talk:Titanic (1997 Film)/Archive 1," Wikipedia, December 13, 2007, http://en.wikipedia.org/?oldid=177757001 (accessed September 30, 2008).

69. Erik Moeller, "Re: [[Talk:Autofellatio]]—Poll," wikien-l, February 13, 2005, http://marc.info/?i=420F062F.50805@gmx.de (accessed February 13, 2005).

70. Wales, "Re: Neo-Nazis to Attack Wikipedia."

71. Jimmy Wales, "Re: Re: A Neo-Nazi Wikipedia," wikien-l, August 27, 2005, http://marc.info/?i=4310D2CB.1060401@wikia.com (accessed August 27, 2005).

72. Jimmy Wales as quoted in Wikimedia, "Meta:Talk:Benevolent Dictator" (oldid=544462).

73. As such, Wales's hybrid leadership style might fit within the "situational school," which advocates different leadership performances as merited by the particular context, see Victor H. Vroom and Philip W. Yetton, *Leadership and Decision-Making* (Pittsburgh: University Of Pittsburgh Press, 1973).

74. Garzarelli and Galoppini, "Capability Coordination in Modular Organization," 36; also see Elinor Ostrom, "Collective Action and the Evolution of Social Norms," *The Journal of Economic Perspectives* 14, no. 3 (Summer 2000): 149.

75. Jimmy Wales, "Where We Are Headed," Foundation-l, June 4, 2006, http://lists.wikimedia.org/pipermail/foundation-l/2006-June/007451.html (accessed August 6, 2008).

76. Ray Saintonge, "Re: Wikimedia Foundation Internal Radio," wikipedia-l, April 12, 2005, http://marc.info/?i=425C3FB4.2070401@telus.net (accessed April 12, 2005).

77. Jimmy Wales, "Re: Harassment Sites," wikien-l, October 24, 2007, http://marc.info/?i=471F6053.4040806@wikia.com (accessed October 24, 2007); Another instance in which Wales is forced to disclaim authority in a message is Jimmy Wales, "Re: Fancruft," wikien-l, July 22, 2006, http://marc.info/?i=44C17378.9030902@wikia.com (accessed July 22, 2006).

78. Jimmy Wales, "Re: Process Wonkery," wikien-l, September 28, 2006, http://marc.info/?i=451BC6B8.8040201@wikia.com (accessed September 28, 2006).

79. Wikimedia, "Benevolent Dictator," Wikimedia, February 2, 2009, http://meta.wikimedia.org/?oldid=1371439 (accessed May 29, 2009).

80. Rick, "Re: Re: Writing About Sexual Topics Responsibly Is Not," wikien-l, February 14, 2005, http://marc.info/?i=20050214062541.66792.qmail@web60609.mail.yahoo.com (accessed February 14, 2005).

Chapter 7: Encyclopedic Anxiety

1. Michael Gorman, "Jabberwiki: The Educational Response, Part II," Britannica Blog: Web 2.0 Forum, June 26, 2007, http://www.britannica.com/blogs/2007/06/jabberwiki-the-educational-response-part-ii/ (accessed June 27, 2007), 5.

2. Wikipedia, "User:Raul654/Raul's Laws," Wikipedia, July 10, 2009, http://en.wikipedia.org/?oldid=301373968 (accessed July 20, 2009).

3. Daniel Headrick, *When Information Came of Age* (Oxford: Oxford University Press, 2000), 145.

4. Henry Hitchens, *Defining the World: The Extraordinary Story of Dr. Johnson's Dictionary* (New York: Farrar, Straus, & Giroux, 2005), 73.

5. For Johnson's original pretense of "fixing" English, see Samuel Johnson, "The Plan of an English Dictionary," in *Oxford Works*, ed. Jack Lynch (2007). "The Plan" originally published in 1747. Johnson's subsequent apology can be seen in Samuel Johnson, "Preface" to *A Dictionary of the English Language*, ed. Jack Lynch (London, 2006), line 84. Originally published in 1755. To review the claims of other lexicographers (i.e., Richard Trench, Ephraim Chambers, and Noah Webster) that their task was to be descriptive historians rather than judgmental critics, see Herbert Charles Morton, *The Story of Webster's Third: Philip Gove's Controversial Dictionary and Its Critics* (New York: Cambridge University Press, 1994), 7; Hitchens, *Defining the World*, 151; Noah Webster, "To John Pickering, December, 1816," in *Letters of Noah Webster*, ed. Harry R. Warfel (New York: Library Publishers, 1953), 357.

6. David Foster Wallace, "Tense Present: Democracy, English, and the Wars Over Usage," *Harper's Magazine* (April 2001).

7. Wikimedia, "Conflicting Wikipedia Philosophies," Wikimedia, 2007, http://meta.wikimedia.org/w/index.php?title=Conflicting_Wikipedia_philosophies& (accessed May 15, 2007).

8. Foster Stockwell, *A History of Information Storage and Retrieval* (Jefferson, NC: Macfarlane, 2001), 133–134; also, see Harvey Einbinder, *The Myth of the Britannica* (New York: Grove Press, 1964), 323–325.

9. Wikipedia, "Wikipedia:Simple English Wikipedia," Wikipedia, October 11, 2008, http://en.wikipedia.org/?oldid=244640674 (accessed November 14, 2008).

10. Anonymous et al., *How to Writer's Guide*, wikiHow, August 21 2007, 2007, http://www.wikihow.com/Writer's-Guide (accessed August 21, 2007).

11. Morton, *The Story of Webster's Third.*

12. Wilson Follett, "Sabotage in Springfield: Webster's Third Edition," *Atlantic Monthly* (January 1962): 74.

13. Jacques Barzun, "The Scholar Cornered: What Is the Dictionary?" *The American Scholar* (Spring 1963): 176.

14. Sheridan Baker, "The Error of Ain't," *College English* 26, no. 2 (November 1964): 91–104; Virginia McDavid, "More on Ain't," *College English* 26, no. 2 (November 1964): 104–105.

15. Morton, *The Story of Webster's Third*, 116, 168, 162.

16. Conservapedia, "Main Page," Conservapedia, July 3, 2008, http://www.conservapedia.com/?oldid=486410 (accessed October 24, 2008).

17. Conservapedia, "Examples of Bias in Wikipedia," Conservapedia, June 2, 2009, http://www.conservapedia.com/?oldid=669839 (accessed June 15, 2009).

18. Lester Haines, "Conservapedia Too Pinko? Try Metapedia," The Register, July 23, 2007, http://www.theregister.co.uk/2007/07/23/metapedia/ (accessed October 24, 2008).

19. Metapedia, "Mission Statement," Metapedia, September 2, 2008, http://en.metapedia.org/?oldid=33062 (accessed October 24, 2008).

20. Tom McArthur, *Worlds of Reference: Lexicography, Learning, and Language from the Clay Tablet to the Computer* (Cambridge, UK: Cambridge University Press, 1986), 67.

21. Johann Heinrich Zedler, *Grosses vollstandiges Universal-Lexicon aller Wissenschaften und Künste,* welche bisshero durch menschlichen Verstand und Witz erfunded und verbessert warden, 64 vols. (Halle and Leipzig: Zedler, 1732-1750); 4-volume supplement (Leipzig: Zedler, 1751-1754), quoted in Headrick, *When Information Came of Age,* 155.

22. Denis Diderot, "The Encyclopedia," in *Rameau's Nephew, and Other Works,* ed. Jacques Barzun and Ralph Henry Bowen (Indianapolis: Hackett Publishing Company, 2001), 295. Paper originally published in 1755.

23. George Gleig, in the preface to the supplement of the third edition as quoted in Britannica, "Encyclopaedia" in *Encyclopedia Britannica,* vol. 15 (1911), 943.

24. Jimmy Wales, quoting Charles Van Doren, in Stacy Schiff, "Know It All: Can Wikipedia Conquer Expertise?" *The New Yorker* (July 31, 2006): 3.

25. Wikipedia, "Charles Van Doren," Wikipedia, September 19, 2007, http://en.wikipedia.org/?oldid=158910101 (accessed September 21, 2007).

26. Herman Kogan, *The Great EB: The Story of the Encyclopaedia Britannica* (Chicago: The University Of Chicago Press, 1958).

27. Einbinder, *The Myth of the Britannica.*

28. Thaddeus K. Oglesby, *Some Truths of History: A Vindication of the South Against the Encyclopedia Britannica and Other Maligners* (Atlanta: The Byrd Printing Company, 1903).

29. Gillian Thomas, *A Position to Command Respect: Women and the Eleventh Britannica* (Metuchen, NJ: The Scarecrow Press, 1992), 5; Britannica, "Ku Klux Klan," in *Encyclopedia Britannica,* vol. 17 (1911), 5; Britannica, "Lynch Law," in *Encyclopedia Britannica,* vol. 15 (1911), 5.

30. Michael McCarthy, "It's Not True about Caligula's Horse; Britannica Checked—Dogged Researchers Answer Some Remarkable Queries," *Wall Street Journal* (April 22, 1999): A1.

31. Joseph McCabe, *The Lies and Fallacies of the Encyclopaedia Britannica: How Powerful and Shameless Clerical Forces Castrated a Famous Work of Reference* (Girard, KS: Haldeman-Julius Publications, 1947).

32. Joseph McCabe, *The Columbia Encyclopedia's Crimes against the Truth* (Girard, KS: Haldeman-Julius Publications, 1951).

33. Wikipedia, "WikiScanner," Wikipedia, September 6, 2007, http://en.wikipedia.org/?oldid=156029747 (accessed September 7, 2007).

34. Robert Darnton, *The Business of Enlightenment: A Publishing History of the Encyclopédie* (Cambridge, MA: The Belknap Press of Harvard University, 1979), 9–13.

35. Stockwell, *A History of Information Storage and Retrieval*, 90.

36. Ibid., 89.

37. Cynthia Koepp, "Making Money: Artisans and Entrepreneurs and Diderot's Encyclopédie," in *Using the Encyclopédie*, ed. Julie Candler Hayes Daniel Brewer (Oxford: Voltaire Foundation, 2002), 138.

38. Einbinder, *The Myth of the Britannica*, 3.

39. Clay Shirky, *Here Comes Everybody: The Power of Organizing without Organizations* (New York: Penguin Press, 2007), 297.

40. Wikipedia, "Criticism of Wikipedia," Wikipedia, March 27, 2007, http://en.wikipedia.org/?oldid=118252212 (accessed March 27, 2007).

41. Wikipedia, "Wikipedia:Replies to Common Objections," Wikipedia, September 20, 2007, http://en.wikipedia.org/?oldid=159141692 (accessed September 27, 2007).

42. Tim O'Reilly, "Web 2.0 Compact Definition: Trying Again," December 10, 2006, http://radar.oreilly.com/archives/2006/12/web-20-compact.html (accessed August 28, 2007); see also Paul Graham, "Web 2.0," November 2005, http://paulgraham.com/web20.html (accessed September 21, 2007); Alex Krupp, "The Four Webs: Web 2.0, Digital Identity, and the Future of Human Interaction," 2006, http://www.alexkrupp.com/fourwebs.html (accessed September 21, 2007).

43. Robert McHenry, "Web 2.0: Hope or Hype?" Britannica Blog: Web 2.0 Forum, June 25, 2007, http://www.britannica.com/blogs/2007/06/web-20-hope-or-hype / (accessed June 27, 2007).

44. Shirky, *Here Comes Everybody*.

45. My understanding is a variation of and response to the "One Smart Guy," "One Best Way," and "One for All" periodization offered by Daniel H. Pink, "The Book Stops Here," *Wired* 13, no. 3 (March 2005).

46. Joseph Reagle, "Four Short Stories about the Reference Work," 2005, http://reagle.org/joseph/2005/refs/4-themes.html (accessed January 12, 2007); Marc Elliott, "Stigmergic Collaboration: The Evolution of Group Work," *M/C Journal* 9, no. 2 (May 2006).

47. Kevin Kelly, *Out of Control: The New Biology of Machines, Social Systems and the Economic World*, reprint edition (Cambridge, MA: Perseus Books Group, 1995).

48. Howard Rheingold, *Smart Mobs* (Cambridge, MA: Perseus Publishing, 2002).

49. James Surowiecki, *The Wisdom of Crowds* (New York: Doubleday, 2004).

50. Michael Gorman, "Jabberwiki: The Educational Response, Part I," Britannica Blog: Web 2.0 Forum, June 25, 2007, http://www.britannica.com/blogs/2007/06/jabberwiki-the-educational-response-part-i/ (accessed June 27, 2007), 4.

51. Michael Gorman, "Web 2.0: The Sleep of Reason, Part II," June 12, 2007, http://www.britannica.com/blogs/2007/06/web-20-the-sleep-of-reason-part-ii/ (accessed June 15, 2007).

52. Jaron Lanier, "Digital Maoism: The Hazards of the New Online Collectivism," Edge, May 30, 2006, http://www.edge.org/documents/archive/edge183.html (accessed June 7, 2006).

53. Andrew Keen, *The Cult of the Amateur: How Today's Internet Is Killing Our Culture* (New York: Doubleday, 2007), 4.

54. Mark Helprin, *Digital Barbarism: A Writer's Manifesto* (New York: Harper, 2009), 18, 41.

55. Lawrence Lessig, "The Solipsist and the Internet (a Review of Helprin's Digital Barbarism)," The Huffington Post, June 15, 2009, http://www.huffingtonpost.com/lawrence-lessig/the-solipsist-and-the-int_b_206021.html (accessed June 15, 2009).

56. Alex Krupp, "Wikipedia, Emergence, and the Wisdom of Crowds," wikipedia-l, May 2, 2005, http://marc.info/?i=4275D17C.6090501@gmail.com (accessed May 2, 2005).

57. Jimmy Wales, "Re: Wikipedia, Emergence, and The Wisdom of Crowds," wikipedia-l, May 2, 2005, http://marc.info/?i=427622EC.8010005@wikia.com (accessed May 2, 2005).

58. Sj, "Re: Wikipedia, Emergence, and The Wisdom of Crowds," wikipedia-l, May 3, 2005, http://marc.info/?i=742dfd06050503151938d6ec8f@mail.gmail.com (accessed May 3, 2005).

59. Joseph Reagle, "Re: Wikipedia, Emergence, and The Wisdom of Crowds," wikipedia-l, May 2, 2005, http://marc.info/?i=200505021101.11598.reagle@mit.edu (accessed May 2, 2005).

60. Yochai Benkler, "Coase's Penguin, or, Linux and the Nature of the Firm," *The Yale Law Journal* 112, no. 3 (2002): 369–446; Lee Sproull, "Online Communities," in *The Internet Encyclopedia*, ed. Hossein Bidgoli (New York: John Wiley, 2003), 733–744.

61. Ward Cunningham, "Keynote: Wikis Then and Now," Wikimania, August 2005, http://commons.wikimedia.org/wiki/Wikimania_2005_Presentations#Keynotes (accessed August 24, 2007).

62. Chris Anderson, *The Long Tail: Why the Future of Business Is Selling Less for More* (New York: Hyperion, 2006); Don Tapscott and Anthony D. Williams, *Wikinomics: How Mass Collaboration Changes Everything* (New York: Portfolio, 2006).

63. Jimmy Wales, "Re: Wikipedia, Emergence, and The Wisdom of Crowds," wikipedia-l, May 3, 2005, http://marc.info/?i=4277C476.3040803@wikia.com (accessed May 3, 2005).

64. Joseph Reagle, "Re: Wikipedia, Emergence, and The Wisdom of Crowds," wikipedia-l, May 4, 2005, http://marc.info/?i=200505040924.52194.reagle@mit.edu (accessed May 4, 2005).

65. Marshall Poe, "The Hive: Can Thousands of Wikipedians Be Wrong? How an Attempt to Build an Online Encyclopedia Touched Off History's Biggest Experiment in Collaborative Knowledge," *Atlantic Monthly* (September 2006).

66. Helprin, *Digital Barbarism*, 65–66.

67. Lanier, "Digital Maoism."

68. Jimmy Wales, "On 'Digital Maoism,'" Edge, May 30, 2006, http://www.edge.org/discourse/digital_maoism.html (accessed June 7, 2006).

69. Yochai Benkler, "On 'Digital Maoism,'" Edge, May 30, 2006, see note 68.

70. Clay Shirky, "Old Revolutions Good, New Revolutions Bad: A Response to Gorman," Many-to-Many, June 13, 2007, http://many.corante.com/archives/2007/06/13/old_revolutions_good_new_revolutions_bad_a_response_to_gorman.php (accessed June 15, 2007), 3.

71. Matthew Battles, "Authority of a New Kind," Britannica Blog: Web 2.0 Forum, June 13, 2007, http://www.britannica.com/blogs/2007/06/authority-of-a-new-kind/ (accessed June 27, 2007).

72. Gorman, "Web 2.0," 2.

73. Richard Stallman, "The Free Universal Encyclopedia and Learning Resource," Free Software Foundation, 1999, http://www.gnu.org/encyclopedia/free-encyclope dia.html (accessed October 11, 2005).

74. Steven Levy, "The Trend Spotter," *Wired* 13, no. 10 (October 2005); Nick Carr, "The Amorality of Web 2.0," Rough Type blog, October 3, 2005, http://www.rought ype.com/archives/2005/10/the_amorality_o.php (accessed October 20, 2005).

75. Michael Gorman, "The Siren Song of the Internet: Part II," Britannica Blog, June 19, 2007, http://www.britannica.com/blogs/2007/06/the-siren-song-of-the-internet -part-ii/ (accessed June 29, 2007).

76. Thomas Mann, "Brave New (Digital) World, Part II: Foolishness 2.0?" Britannica Blog: Web 2.0 Forum, June 27, 2007, http://www.britannica.com/blogs/2007/06 /brave-new-digital-world-part-ii-foolishness-20/ (accessed June 27, 2007).

77. Adrian Johns, "The Birth of Scientific Reading," *Nature* 409, no. 287 (January 18, 2001): 633.

78. Glenn Reynolds, *An Army of Davids: How Markets and Technology Empower Ordinary People to Beat Big Media, Big Government, and Other Goliaths* (Nashville, TN: Nelson Current, 2006).

79. Anderson, *The Long Tail.*

80. Battles, "Authority of a New Kind."

81. Shirky, "Old Revolutions Good, New Revolutions Bad."

82. Battles, "Authority of a New Kind," 2.

83. danah boyd, "Knowledge Access as a Public Good," Britannica Blog: Web 2.0 Forum, June 27, 2007, http://www.britannica.com/blogs/2007/06/knowledge-access -as-a-public-good/ (accessed June 27, 2007).

84. danah boyd, "Wikipedia, Academia and Seigenthaler," Many2Many, December 17, 2005, http://many.corante.com/archives/2005/12/17/wikipedia_academia_and _seigenthaler.php (accessed December 17, 2005).

85. Peter Burke, *A Social History of Knowledge: From Gutenberg to Diderot* (Cambridge, UK: Polity Press, 2000).

86. On communism, see Andrew Keen, "Web 2.0: The Second Generation of the Internet Has Arrived. It's Worse Than You Think." *The Weekly Standard* (February 15, 2006); on socialism, see Richard Stallman, "Bill Gates and Other Communists," CNET News.com, February 15, 2005, http://news.cnet.com/Bill-Gates-and-other -communists/2010-1071_3-5576230.html (accessed September 15, 2007); Richard Stallman, "Pigdog Interviews Richard Stallman," GNUisance, February 15, 2005, http://www.pigdog.org/interviews/stallman/interview_with_stallman3.html (accessed September 15, 2007); Larry Lessig, "Et Tu, KK? (Aka, No, Kevin, This Is Not

'Socialism')," Lessig Blog, May 28, 2009, http://www.lessig.org/blog/2009/05/et_tu_kk_aka_no_kevin_this_is.html (accessed May 29, 2009); on Maoism, see Lanier, "Digital Maoism."

87. Hayeck is also the central theorist in Cass Sunstein, *Infotopia: How Many Minds Produce Knowledge* (Oxford: Oxford University Press, 2006).

88. Mann, "Brave New (Digital) World, Part II," 4.

89. Michael Gorman, "Revenge of the Blog People!" *Library Journal* (February 15, 2005).

90. Wales is described as an "Enlightenment kind of guy" in Schiff, "Know It All," 3; he is described as a "counter-enlightenment guy" in Keen, *The Cult of the Amateur*, 41.

91. Suzanne Briet, "What Is Documentation? (Qu'Est-Ce Que La Documentation?, 1951)," in *What is Documentation: English Translation of the Classic French Text*, ed. Ronald E. Day and Laurent Martinet with Hermina G. B. Anghelescu (trans.) (Lanham, MD: Scarecrow Press, 2006), 42.

92. Pantisocracy was never realized and after *Metropolitana*'s initial publication, with editorial changes not to his liking, Coleridge withdrew from the project and it subsequently failed. However, his introduction expounding upon its "method" of organizing knowledge according to a progression of intellectual disciplines, rather than alphabetically, would influence other encyclopedists. For more on Coleridge's projects see McArthur, *Worlds of Reference*, 157; Stockwell, *A History of Information Storage and Retrieval*, 109.

93. Simon Winchester, *The Professor and the Madman* (New York: HarperCollins, 1998), 38.

94. Ibid.

95. McArthur, *Worlds of Reference*, 93.

96. Caius Plinius Cecilius, "To Baebius Macer," Ancient History Sourcebook, 1998, http://www.fordham.edu/halsall/ancient/pliny-letters.html (accessed March 18, 2005), letter XXVII. Originally published circa 100 CE.

97. Wikipedia, "Wikipedia:Wikipediholic," Wikipedia, May 10, 2009, http://en.wikipedia.org/?oldid=289128389 (accessed May 29, 2009); Wikipedia, "Wikipedia:Editcountitis," Wikipedia, April 30, 2009, http://en.wikipedia.org/?oldid=287150522 (accessed May 29, 2009).

98. Wikimedia, "Elections for the Board of Trustees of the Wikimedia Foundation, 2006/En," Wikimedia, September 23, 2006, http://meta.wikimedia.org/?oldid=439005 (accessed December 4, 2006); Wikimedia, "Board Elections/2008/En," Wikimedia, June 27, 2008, http://meta.wikimedia.org/?oldid=1059669 (accessed July 30, 2009).

99. Wikipedia, "Wikipedia:Deceased Wikipedians," Wikipedia, May 28, 2009, http://en.wikipedia.org/?oldid=292924929 (accessed May 29, 2009).

100. Johnson, "Preface."

101. A. J. Jacobs, *The-Know-It-All: One Man's Humble Quest to Become the Smartest Person in the World* (New York: Simon & Schuster, 2004), 88.

102. holotone, "Wikipedia Showdown!" grupthink, August 3, 2006, http://www.grupthink.com/topic/364 (accessed August 3, 2006).

103. Oded Nov, "What Motivates Wikipedians?" *Communications of the ACM* 50, no. 11 (November 2007): 60-64.

104. Alexandra Shimo, "Prolific Canadian Is King of Wikipedia," *The Globe and Mail* (August 4, 2006).

105. Wikipedia, "User:Seth Ilys/Dot Project," Wikipedia, August 4, 2006, http://en.wikipedia.org/?oldid=67572849 (accessed August 22, 2006).

106. Wikipedia, "Wikipedia:Blocking Policy," Wikipedia, April 6, 2006, http://en.wikipedia.org/?oldid=47260197 (accessed April 6, 2006).

107. Nicolas A. Basbanes, *A Gentle Madness: Bibliophiles, Bibliomanes, and the Eternal Passion for Books* (New York: Henry Holt & Company, 1999).

108. Charles Arthur, "Log on and Join in, but Beware the Web Cults," *The Guardian* (December 15, 2005).

109. Gorman, "Jabberwiki."

110. Helprin, *Digital Barbarism*, 18.

111. Lanier, "Digital Maoism."

112. Andrew Orlowski, "Wikipedia Founder Admits to Serious Quality Problems," The Register, October 18, 2005, http://www.theregister.co.uk/2005/10/18/wikipedia_quality_problem/ (accessed October 20, 2005).

113. Paulo Correa, Alexandra Correa, and Malgosia Askanas, "Wikipedia: A Techno-Cult of Ignorance," Aetherometry, December 2005, http://www.aetherometry.com/Electronic_Publications/Politics_of_Science/Antiwikipedia/awp_index.html (accessed April 17, 2009).

114. Paul Otlet, "Transformations in the Bibliographical Apparatus of the Sciences," in *International Organization and Dissemination of Knowledge: Selected Essays of Paul Otlet*, ed. W. Boyd Rayward (Amsterdam: Elsevier, 1990), 149. Paper originally published in 1918.

115. Kathy Bowrey and Matthew, "Rip, Mix, Burn," *First Monday* 7, no. 8 (July 2005).

116. Michael Gorman, "Google and God's Mind: The Problem Is, Information Isn't Knowledge," *Los Angeles Times* (December 17, 2004).

117. Kevin Kelly, "Scan This Book! What Will Happen to Books? Reader, Take Heart! Publisher, Be Very, Very Afraid. Internet Search Engines Will Set Them Free. A Manifesto," *The New York Times Magazine* (May 14, 2006): 2.

118. Ibid., 2–3.

119. Gorman, "Google and God's Mind," 2.

120. Keen, *The Cult of the Amateur*, 57.

121. Kevin Drum, "Google and the Human Spirit: A Reply to Michael Gorman," *Washington Monthly* (December 17, 2004).

122. Paul Otlet, "The Science of Bibliography and Documentation," in *International Organization and Dissemination of Knowledge: Selected Essays of Paul Otlet*, ed. W. Boyd Rayward (Amsterdam: Elsevier, 1990), 79. Paper originally published in 1903.

123. Johns, "The Birth of Scientific Reading"; Ann Blair, "Reading Strategies for Coping with Information Overload Ca. 1550–1700," *Journal of the History of Ideas* 64, no. 1 (2003): 11–28.

124. Francis Bacon, "Of Studies," in *Bacon's Essays; With Introduction, Notes, and Index*, ed. Edwin A. Abbott (London: Longman's, 1879), 185.

125. Hugh Of St. Victor, "The Seven Liberal Arts: On Study and Teaching," in *The Portable Medieval Reader*, ed. James Bruce Ross and Mary Martin McLaughlin, Viking Portable Library (Penguin, 1977), 584–585. Essay originally published in the twelfth century.

126. Gorman, "Revenge of the Blog People!" Much as Godwin's Law predicts an unfavorable Nazi analogy in a long discussion, arguments about technology inevitably prompt a comparison with the Luddites, see Michael Gorman, "The Siren Song of the Internet: Part I," Britannica Blog, June 18, 2007, http://www.britannica.com/blogs/2007/06/the-siren-song-of-the-internet-part-i/ (accessed June 29, 2007); Clay Shirky, "Gorman, Redux: the Siren Song of the Internet," June 20, 2007, http://many.corante.com/archives/2007/06/20/gorman_redux_the_siren_song_of_the_internet.php (accessed June 29, 2007).

127. Andrew Orlowski, "There's No Wikipedia Entry for 'Moral Responsibility,'" The Register, December 13, 2005, http://www.theregister.co.uk/2005/12/12/wikipedia_no_responsibility/ (accessed December 13, 2005).

128. Nicholas Carr, "Stabbing Polonius," Rough Type, April 26, 2007, http://www.roughtype.com/archives/2007/04/sanger_1.php (accessed April 30, 2007).

129. Lanier, "Digital Maoism."

130. Gorman, "Web 2.0."

131. Helprin, *Digital Barbarism*, 65.

132. Carr, "Stabbing Polonius."

133. Wikipedia, "Wikipedia:Verifiability," Wikipedia, November 14, 2008, http://en.wikipedia.org/?oldid=251829388 (accessed November 14, 2008).

134. Tapscott and Williams, *Wikinomics*.

135. Cass R. Sunstein, *Why Societies Need Dissent* (Cambridge, MA: Harvard University Press, 2003).

136. Jeremy Wagstaff, "The New Cliche: 'It's the Wikipedia of . . .'" Loose Wire, September 29, 2005, http://loosewire.typepad.com/blog/2005/09/the_new_cliche_.html (accessed October 2, 2005).

137. Larry Sanger, "The Early History of Nupedia and Wikipedia: A Memoir," Slashdot, April 18, 2005, http://features.slashdot.org/article.pl?sid=05/04/18/164213 (accessed April 19, 2005); PeopleProjectsAndPatterns, "Wikipedia," Cunningham & Cunningham, 2007, http://www.c2.com/cgi/wiki?WikiPedia (accessed September 21, 2007); danah boyd, "Academia and Wikipedia," Many2Many, January 4, 2005, http://many.corante.com/archives/2005/01/04/academia_and_wikipedia.php (accessed January 4, 2005); Clay Shirky, "Wikipedia: Me on Boyd on Sanger on Wales," Many2Many, January 5, 2005, http://many.corante.com/archives/2005/01/05/wikipedia_me_on_boyd_on_sanger_on_wales.php (accessed January 5, 2005).

138. Larry Sanger, "Britannica or Nupedia? The Future of Free Encyclopedias," Kuro5hin, July 25, 2001, http://www.kuro5hin.org/story/2001/7/25/103136/121 (accessed June 6, 2006).

139. Peter Jacso, "Peter's Picks & Pans," *Online* 26 (Mar/Apr 2002): 79–83.

140. Wikimedia Foundation, "Wikipedia Reaches 2 Million Articles," Wikimedia Foundation, September 13, 2007, http://wikimediafoundation.org/wiki/Wikipedia_Reaches_2_Million_Articles (accessed September 13, 2007); Wikipedia, "Wikipedia:Million Pool," Wikipedia, August 5, 2007, http://en.wikipedia.org/?oldid=149380521 (accessed September 7, 2007).

141. Burt Helm, "Wikipedia: 'A Work in Progress,'" Business Week Online, December 14, 2005, http://www.businessweek.com/technology/content/dec2005/tc20051214_441708.htm (accessed December 15, 2005).

142. Wikipedia, "Wikipedia:What Wikipedia Is Not," Wikipedia, May 28, 2009, http://en.wikipedia.org/?oldid=292975573 (accessed May 29, 2009).

143. Wikipedia, "Wikipedia:10 Things You did Not Know About Wikipedia," Wikipedia, September 3, 2007, http://en.wikipedia.org/?oldid=155431119 (accessed September 7, 2007).

144. The Guardian, "In Praise of . . . the Wikipedia," *The Guardian* (December 9, 2005).

145. John Quiggin, "Wikipedia and Sausages," Out Of the Crooked Timer, March 1, 2006, http://crookedtimber.org/2006/03/01/wikipedia-and-sausages/ (accessed March 1, 2006).

146. Bill Thompson, "Wikipedia—A Flawed and Incomplete Testament to the Essential Fallibility of Human Nature," BBC's Digital Revolution Blog, July 23, 2009, http://www.bbc.co.uk/blogs/digitalrevolution/2009/07/wikipedia.shtml (accessed July 23, 2009).

147. Benkler, "On 'Digital Maoism.'"

148. Cory Doctorow, "On 'Digital Maoism,'" Edge, May 30, 2006, see note 68.

149. Kevin Kelly, "On 'Digital Maoism,'" Edge, May 30, 2006, see note 68.

150. Jim Giles, "Internet Encyclopaedias Go Head to Head," *Nature* (December 14, 2005).

151. Nate Anderson, "Britannica Attacks Nature in Newspaper Ads," Ars Technica, April 5, 2006, http://arstechnica.com/news.ars/post/20060405-6530.html (accessed April 5, 2006); Wikipedia, "Wikipedia:External Peer Review/Nature December 2005 /Errors," Wikipedia, February 9, 2006, http://en.wikipedia.org/?oldid=38886868 (accessed April 6, 2006).

152. Orlowski, "Wikipedia Founder Admits to Serious Quality Problems."

153. Robert McHenry, quoted in Schiff, "Know It All," 7.

154. Carr, "The Amorality of Web 2.0."

155. Peter Denning, Jim Horning, David Parnas, and Lauren Weinstein, "Inside Risks: Wikipedia Risks," *Communications of the ACM* 48, no. 12 (2005): 152; Keen, *The Cult of the Amateur*, 27.

156. Wikipedia, "Every Time You Masturbate . . . God Kills a Kitten," Wikipedia, September 11, 2007, http://en.wikipedia.org/?oldid=157104187 (accessed September 13, 2007); Keen, *The Cult of the Amateur*, 29.

157. Nick Carr, "Is Google Making Us Stupid?" *Atlantic Monthly* (July 2008); a well-researched and persuasive argument of detrimental media effects can be found in Mark Bauerlein, *The Dumbest Generation: How the Digital Age Is Stupefied as Young Americans and Jeopardizes Our Future: Or, Don't Trust Anyone Under 30* (New York: Tarcher/Penguin, 2008).

158. Gorman, "Web 2.0."

159. Gorman, "Jabberwiki."

160. Shirky, "Old Revolutions Good, New Revolutions Bad."

161. Scott Mclemee, "Mass Culture 2.0," Inside Higher Ed, June 20, 2007, http://insidehighered.com/views/2007/06/20/mclemee (accessed June 27, 2007), 4.

162. Douglas Adams, "How to Stop Worrying and Learn to Love the Internet," *The Sunday Times* (August 29, 1999).

163. Clay Shirky, "Newspapers and Thinking the Unthinkable," Shirky, March 13, 2009, http://www.shirky.com/weblog/2009/03/newspapers-and-thinking-the-unthinkable/ (accessed March 14, 2009); Steve Weber, *The Success of Open Source* (Cambridge, MA: Harvard University Press, 2004).

164. Hunter R. Rawlings III, "Information, Knowledge, Authority, and Democracy" (ARL Keynote), October 10, 2007, http://www.arl.org/bm~doc/mm-f07-rawlings.pdf (accessed April 17, 2009).

Chapter 8: Conclusion: "A Globe in Accord"

1. Conversely, although Wells would have been fascinated with the "technological paraphernalia of our networked age," "he would equally have cared little for its individualism, cultural relativism and a lack of respect for professionals and experts," as argued by Dave Muddiman, "The Universal Library as Modern Utopia: The Information Society of H. G. Wells," *Library History* 14 (November 1998): 98.

2. Coincidentally, Mike Godwin, author of Godwin's Law and the seminal "Nine Principles for Making Virtual Communities Work," joined the Wikimedia Foundation as its general counsel and legal coordinator in July 2007, as announced by Florence Devouard, "Welcome Mike !" Foundation-l, July 3, 2007, http://lists.wikimedia.org/pipermail/foundation-l/2007-July/031128.html (accessed July 24, 2007). Wikipedians expressed hope that his insight and experience with online community would help Wikipedia address some of its challenges.

3. Carol Tavris and Elliot Aronson, *Mistakes Were Made (But Not by Me): Why We Justify Foolish Beliefs, Bad Decisions, and Hurtful Acts* (New York: Harcourt, 2007).

4. Jimmy Wales, "A Culture of Co-Operation," wikipedia-l, November 5, 2001, http://marc.info/?l=wikipedia-l&m=104216623606518&w=2 (accessed August 23, 2005).

5. Stacy Schiff, "Know It All: Can Wikipedia Conquer Expertise?" *The New Yorker* (July 31, 2006): 1.

6. Abraham Lincoln, "First Inaugural Address," March 4, 1861, http://www.bartleby.com/124/pres31.html (accessed November 14, 2007).

7. Peter Kropotkin, "Mutual Aid: A Factor of Evolution," 1902, http://dwardmac.pitzer.edu/Anarchist_Archives/kropotkin/mutaidcontents.html (accessed December 20, 2007).

8. Larry Sanger, quoted in Wade Roush, "Larry Sanger's Knowledge Free-for-All," *Technology Review* (January 2005): 21.

9. Peter Kollock, "The Economies of Online Cooperation: Gifts and Public Goods in Cyberspace," in *Communities in Cyberspace,* ed. Marc Smith and Peter Kollock (London: Routledge Press, 1999), 220.

10. Larry Sanger, "The Early History of Nupedia and Wikipedia: A Memoir," Slashdot, April 18, 2005, http://features.slashdot.org/article.pl?sid=05/04/18/164213 (accessed April 19, 2005).

11. Larry Sanger, "Text and Collaboration, Part I," Kuro5hin, May 28, 2006, http://www.kuro5hin.org/story/2006/5/26/162017/011 (accessed June 8, 2006), 4.

12. Ibid., 6.

13. Stephen Foley, "So Is Wikipedia Cracking Up?" *The Independent* (February 3, 2009).

14. Eric Goldman, "Wikipedia Will Fail in Four Years," Technology & Marketing Law Blog, December 5, 2006, http://blog.ericgoldman.org/archives/2006/12/wikipedia_will_1.htm (accessed February 18, 2009); Nate Anderson, "Doomed: Why Wikipedia Will Fail," Ars Technica, February 12, 2009, http://arstechnica.com/web/news/2009/02/doomed-why-wikipedia-will-fail.ars (accessed February 12, 2009).

15. Wikipedia, "User:Raul654/Raul's Laws," Wikipedia, July 10, 2009, http://en.wikipedia.org/?oldid=301373968 (accessed July 20, 2009).

16. Max Weber, *Economy and Society: An Outline of Interpretive Sociology,* vol. 1, ed. Claus Wittich Guenther Roth (Berkeley: University of California Press, 1978), 244. Originally published in 1914.

Index